PLANT SCIENCE RESEARCH AND PRACTICES

CYANOBACTERIA

ECOLOGICAL IMPORTANCE, BIOTECHNOLOGICAL USES AND RISK MANAGEMENT

PLANT SCIENCE RESEARCH AND PRACTICES

Additional books in this series can be found on Nova's website under the Series tab.

Additional e-books in this series can be found on Nova's website under the e-book tab.

PLANT SCIENCE RESEARCH AND PRACTICES

CYANOBACTERIA

ECOLOGICAL IMPORTANCE, BIOTECHNOLOGICAL USES AND RISK MANAGEMENT

DOUGLAS DAVISON
EDITOR

nova publishers
New York

Library of Congress Cataloging-in-Publication Data

Cyanobacteria : ecological importance, biotechnological uses and risk management / editor: Douglas Davison.
pages cm. -- (Plant science research and practices)
Includes index.
ISBN 978-1-61470-925-1 (hardcover)
1. Cyanobacteria. I. Davison, Douglas, editor. II. Series: Plant science research and practices.
QR99.63.C919 2014
579.3'9--dc23

2014036532

Published by Nova Science Publishers, Inc. † New York

CONTENTS

PREFACE

Cyanobacteria are ubiquitous photoautotrophic prokaryotes commonly found in aquatic and terrestrial ecosystems. Some can even thrive under immense oxidative stress in extreme environments, such as hot/cold deserts and hot springs. Formation of reactive oxygen species (ROS) in cyanobacteria not only occurs in the respiratory metabolism, but also involved in the receptor–side of the electron transport chain during photosynthesis. This book discusses the ecological importance of cyanobacteria. It also discusses some biotechnological uses and risk management.

Chapter 1 – When the photosynthetic pigments like chlorophyll *a*, carotenoids and phycobiliproteins are irradiated with sunlight, ROS are inevitably produced via the interaction among the excited pigments, adjacent molecules and molecular oxygen. To protect themselves against the oxidative stress, cyanobacteria have developed efficient defence systems. The most important defence mechanisms are (1) the unique cyanobacterial orange carotenoid protein and iron stress induced protein (IsiA) that dissipate the excessive light energy as heat, (2) the protective exopolysaccharide sheath containing UV absorbing pigments like scytonemin and mycosporine–like amino acids, (3) antioxidant enzymes such as SODs, catalases and peroxidases, (4) nonenzymatic antioxidants like carotenoids, ascorbate, α–tocopherol and glutathione, (5) D1 protein turnover and PSII repair, and (6) small CAB–like proteins and high light inducible polypeptides that stabilize PSI and PSII against oxidative stress. In spite of these numerous protective tools, the impact of ROS on cyanobacteria is evidenced by the inhibition of photosynthesis, bleaching of photosynthetic pigments, DNA breakage, degradation of membrane lipids, proteins and thylakoids, and programmed cell death. On a macroscopic scale, the environmental impact of ROS formation in cyanobacteria is directly visible as the whitening of the symbiotic coral reefs in the ocean, and the seasonal retreat of algal blooms in waters. The cytotoxic effects of ROS on cyanobacteria can be positively exploited for water purification whereas the combination of photoinhibition and oxidative stress may find application in the conservation of underground cultural heritage threatened by cyanobacterial attack.

Chapter 2 – This review focuses on cyanobacteria which are copiously studied in the last two decades with respect to their presence and activity in petroleum hydrocarbon or its derivates (PH) polluted environments. The ability of consortia of (selected) cyanobacteria and hydrocarbonoclastic heterotrophic bacteria (HCHB) to degrade PH is clearly demonstrated by an increasing number of scientific reports that strongly argue that these mixed populations can be very important participants in the bioremediation of PH polluted sites. There are clear results which show that HCHB work better together with cyanobacteria for the bioremediation of PH contaminated sites, the general view being that the degradation of PH is carried out predominantly by HCHB, the contribution of cyanobacteria being mostly indirect. This is an important contribution mainly :i) by providing the molecular oxygen needed for aerobic respiration of heterotrophic bacteria during oxic degradation of PH; ii) by producing and excreting organic substances which sustain the activity of HCHB and iii) by offering HCHB surfaces to attach to. Some future prospects are also presented, mostly form a microbiological perspective, concerning the strategies to enhance petroleum hydrocarbon degradation by increasing the basic knowledge about the biology of cyanobacteria at molecular, cellular/population and ecological level, and integration with other cyanobacteria-based biotechnologies.

Chapter 3 – Eutrophication is a global and serious problem in lakes and reservoirs all over the world. An obvious and risky symptom of eutrophication is the rapid growth and accumulation of phytoplankton (called blooms) leading to discoloration of the affected waters. Blooms are a prime agent of the water quality deterioration, deoxygenation of bottom waters (hypoxia and anoxia) and toxicity. Toxins produced by blooms can adversely affect animals and human health in waters used for recreational and drinking purposes. Numerous genera are able to form blooms. However, cyanobacteria are the most notorious blooms formers: they are adept at exploiting nutrient-enriched conditions. One of their major problems arises from their ability to produce toxins. Consequently, preventing eutrophication is an important issue in lakes and reservoir management. In order to develop an eutrophication (or cyanobacteria) prevention system, a good starting point can be the diagnosis of the situation in the area where the selected reservoirs in this study are located. The Trasona reservoir was used as a pilot reservoir in which a more exhaustive study was carried out: many chemical, physico–chemical and biological parameters were monitored for years in it, mathematical models were developed to forecast eutrophication (and cyanobacteria presence), and therefore the risk of cyanotoxins, etc. The aim of this work was to obtain predictive models able to perform an early detection of cyanotoxins by using as predictors different chemical, physico–chemical and biological measurements. The study highlighted the importance of the different kinds of cyanobacteria in cyanotoxins production by using the above-mentioned models. Moreover, the models obtained were tried in other fifteen Cantabrian Basin reservoirs to illustrate that they could be used to successfully predict the cyanobacteria presence in the reservoirs of the study area. Finally, conclusions of this innovative research are exposed.

Chapter 4 – Freshwater algal blooms have become a growing concern in Macau Storage Reservoir (MSR), which were caused by a high level of cyanobacteria, particularly Microcystis spp. and C. raciborskii that can produce microcystin and cylindrospermopsin, respectively. Long-time exposure to these cyanotoxins may affect public health, thus reliable detection and quantification of the algae species were challenging in water quality management. The aim of this study was to develop accurate and sensitive molecular methods on detecting cyanobacterial species and cyanotoxins-producing genotypes. The cyanobacteria, Microcystis spp. and C. raciborskii were identified and quantified by polymerase chain reaction (PCR), multiplex PCR and real-time quantitative PCR (qPCR) techniques. As well, the cyanotoxins-producing genotypes were also analyzed by PCR and qPCR using cylindrospermopsin polyketide synthetase gene (pks) and microcystin synthetase genes (mcys). The related water eutrophication, phytoplankton community diversity were measured accordingly. It was found out that TSI result of 65-82, indicated that MSR was categorized as a eutrophic-hypereutrophic reservoir, with the dominance of Cyanophyta in 2011, and of Chlorophyta and Bacillariophyta in 2012. The PCR (including multiplex PCR) results showed that the techniques were successful for identifying cyanobacterial species and cyanotoxins-producing genotypes in pure cultures (or plasmid), mixed cultures, and water samples in MSR. While qPCR results were proved to be applied in quantifying the cell number of cyanobacteria, Microcystis spp. and C. raciborskii, as well as the gene copy nember of microcystin- and cylindrospermopsin-producing genotypes. When the target species above 1 million cells/L, similar cell numbers estimated by microscopic counting and qPCR were obtained. Further quantification in water samples indicated that the ratio of cells number estimated by microscopy and by qPCR was 0.4-12.9 for cyanobacteria and 0.2-3.9 for C. raciborskii. However, Microcystis spp. was not observed by manual counting, while it can be detected at low levels by qPCR, suggesting that qPCR is more sensitive and accurate than microscopic counting. On the aspect of cyanotoxins, there was a strong correspondence between the presence of the pks gene numbers and cylindrospermopsin concentrations (R2=0.95) determined by HPLC, while weak correlations were obtained between the mcys gene numbers and microcystin concentrations. Furthermore, the pks gene numbers were strongly related to Cylindrospermopsis (R2=0.88), cyanobacterial cell numbers (R2=0.96), total algae numbers (R2=0.95) and chlorophyll-a concentrations (R2=0.83), consistent with the dominant species of Cylindrospermopsis among the cyanobacteria existing in MSR. The water quality parameters NH4-N (R2=0.68) and pH (R2=0.89) were most highly correlated with the pks gene numbers.

Chapter 5 – Microcystins (MCs) are secondary metabolites produced by freshwater cyanobacteria that have been associated with severe episodes of human and animal acute hepatotoxicity. MC is produced by several cyanobacteria species from t different genera such as *Microcystis, Anabaena, Planktothrix.* These have worldwide distribution in freshwater bodies and are known for their hepatotoxic effects in mammals by inhibiting protein phosphatase. So far, concern is growing over MCs contamination in drinking water in many countries around the world. MC is recalcitrant to conventional water treatment and is often detected even after chemical treatment of waters. Chemicals used to kill the algae are potentially toxic and may remain in the water to pose greater danger when consumed. Consequently, the biological methods of removing MCs are an alternative to the disadvantages of using chemicals. In this chapter, a review of the literature on the bacterial degradation of MCs was performed. Methods of isolation of bacteria were collected and a review of the state of knowledge of the subject was made including degradation mechanisms. Furthermore, the authors have exposed the advantages and disadvantages of biological methods and their potential application to water treatment biotechnology.

Chapter 6 – The global climate change with an increase in solar ultraviolet (UV) radiations on Earth's surface has generated tremendous concern about its negative impact on all sun-exposed photosynthetic life-forms including cyanobacteria. UV (280 - 400 nm) can affect a number of physiological and biochemical processes in cyanobacteria either directly or indirectly by the generation of reactive oxygen species. Drastic effects caused by UV radiation may imbalance the entire ecosystems. However, many cyanobacteria are able to develop several defense mechanisms against the damaging effects imposed by increased UV radiation. These strategies include the restoration of genetic material by means of excision repair and photoreactivation, repair and re-synthesis of D1 and D2 proteins of PSII reaction center, activation of antioxidant systems, and biosynthesis of several photoprotectants such as mycosporine-like amino acids, scytonemin, carotenoids and polyamines. This chapter presents an overview on the current knowledge of the physiological as well as biochemical adaptation of cyanobacteria with regard to the function of potent defense mechanisms in response to intense solar UV radiation.

Chapter 7 – An easy-to-operate pilot bioreactor designed for biogas H_2S and CO_2 adsorption has successfully upgraded biogas produced from the excreta of 1500 pigs on a farm in rural Taiwan. The bioreactor also seals against unpleasant odors. It costs less than US$200 and can handle at least 3 hours of continuous biogas upgrading, thus contributing to the household gas energy capacity for pig farmers. The bioreactor is designed to dispense a volume of controlled biogas through a small aperture micropore dispenser. The small biogas bubbles pass through water containing cyanobacteria, dissolving the biogas's H_2S and CO_2 into the water, which are then absorbed by cyanobacteria. Key aspects of the design are the amount of water and types of cyanobacteria in the tank, as well as the path length of the bubbles rising through the water. After treatment, the concentration of H_2S in the biogas dropped from 3040 ppm to 2 ppm at the first 10 minutes, 6 ppm after one hour, and arose to 13 ppm after 3 hours. It was acceptably within the upper limit of the threshold for the harmful effects of natural gas (50 ppm). The bioreactor's efficiency could be improved with a longer container for containing the cyanobacteria and a smaller micropore size. The quality of the original biogas could also be upgraded. This system could also lower the CO_2 concentration from 27% to 14% after running for 3 hours.

Chapter 8 – Cyanobacteria are photosynthetic bacteria that have played important roles in the development of the Earth and its atmosphere. Accordingly, novel experimental techniques for modifying their regulation and function are of great interest. Cell-penetrating peptides (CPPs) have attracted scientists' attention in recent decades, because they can overcome the hydrophobic plasma membrane barrier and enter cells directly. This phenomenon is known as protein transduction. In this study, the authors demonstrate that one CPP, the nona-arginine (R9) peptide, is able to deliver noncovalently associated green fluorescent protein (GFP) into *Synechococcus elongatus* PCC 7942 cyanobacteria. Reduction of GFP internalization by physical and pharmacological inhibitors demonstrated that uptake of uncomplexed GFP by cyanobacteria is mediated by classical endocytosis. CPP-mediated protein transduction of R9/GFP complexes was inhibited by treatment with N-ethylmaleimide (NEM), and macropinocytic inhibitors diminished CPP-mediated protein transduction of R9/GFP complexes in either the absence or presence of NEM. These two lines of evidence suggest that the major mechanism for CPP-mediated protein transduction is macropinocytosis, and classical endocytosis plays a minor role. Collectively, the authors' CPP investigations offer the insight of the understanding of protein import in cyanobacteria.

In: Cyanobacteria
Editor: Douglas Davison
ISBN: 978-1-61470-925-1

Chapter 1

REACTIVE OXYGEN SPECIES IN CYANOBACTERIA

***Paul Hsieh*[1*] *and Jens Z. Pedersen*[2]**
[1]Laboratory of Biology of Algae, Department of Biology,
University of Rome 'Tor Vergata' Rome, Italy
[2]EPR Laboratory, Department of Biology,
University of Rome 'Tor Vergata', Rome, Italy

ABSTRACT

Cyanobacteria are ubiquitous photoautotrophic prokaryotes commonly found in aquatic and terrestrial ecosystems. Some can even thrive under immense oxidative stress in extreme environments, such as hot/cold deserts and hot springs. Formation of reactive oxygen species (ROS) in cyanobacteria not only occurs in the respiratory metabolism, but also at the receptor–side of the electron transport chain during photosynthesis. Additionally, when the photosynthetic pigments like chlorophyll *a*, carotenoids and phycobiliproteins are irradiated with sunlight, ROS are inevitably produced via the interaction among the excited pigments, adjacent molecules and molecular oxygen. To protect themselves against the oxidative stress, cyanobacteria have developed efficient defence systems. The most important defence mechanisms are (1) the unique cyanobacterial orange carotenoid protein and iron stress induced protein (IsiA) that dissipate the excessive light energy as heat, (2) the protective exopolysaccharide sheath containing UV absorbing pigments like scytonemin and mycosporine–like amino acids, (3) antioxidant enzymes such as SODs, catalases and peroxidases, (4) nonenzymatic antioxidants like carotenoids, ascorbate, α–tocopherol and glutathione, (5) D1 protein turnover and PSII repair, and (6) small CAB–like proteins and high light inducible polypeptides that stabilize PSI and PSII against oxidative stress. In spite of these numerous protective tools, the impact of ROS on cyanobacteria is evidenced by the inhibition of photosynthesis, bleaching of photosynthetic pigments, DNA breakage, degradation of membrane lipids, proteins and thylakoids, and programmed cell death. On a macroscopic scale, the environmental impact of ROS formation in cyanobacteria is directly visible as the whitening of the symbiotic coral reefs in the ocean, and the seasonal retreat of algal blooms in waters. The cytotoxic effects of ROS on cyanobacteria can be positively exploited for water purification whereas the combination of

* Corresponding author: Email: goldenage@mail2000.com.tw.

photoinhibition and oxidative stress may find application in the conservation of underground cultural heritage threatened by cyanobacterial attack.

1. INTRODUCTION

Between 3.2 and 2.4 billion years ago, the Earth's atmosphere became oxidating partly due to the proliferation of cyanobacteria (also known as blue–green algae) able to carry out oxygen–producing photosynthesis (Brocks et al., 1999). The accumulation of O_2 promoted the development of aerobic organisms that used O_2 as a powerful electron acceptor. Today, cyanobacteria account for 20–30% of the Earth's total photosynthetic productivity. They convert solar energy into biomass–stored chemical energy at a rate around 450 TW (Waterbury et al., 1979). Despite being one of the Earth's primary producers, cyanobacteria constantly need to deal with the damaging effects of reactive oxygen species (ROS) inevitably generated during photosynthesis and respiration.

The most important ROS produced in cyanobacteria are singlet oxygen ($^1O_2$), the hydroxyl radical (•OH), superoxide ($\bullet O_2^-$) and hydrogen peroxide (H_2O_2); each of them possesses distinct chemical and physical properties that lead to different reactivity, toxicity and targets. $^1O_2$ has half–life of ~2 μs in water (Wilkinson et al., 1995) that corresponds to a diffusion distance of ~220 nm and a 5% remaining concentration of $^1O_2$ for a period of 3τ (Redmond and Kochevar, 2006). $^1O_2$ can be quenched by other compounds to produce excited complexes, either through energy dissipation when energy is transferred to the ground–state $^3O_2$ and dissipated as heat, or through an oxidation reaction. Under physiological conditions $^1O_2$ can react with most unsaturated compounds to produce a variety of peroxides. $^1O_2$ in cyanobacteria is reactive with various compounds, including unsaturated fatty acids in membrane lipids (Triantaphylidès et al., 2008), side–chains of the amino acids Trp, Tyr, His, Met and Cys in proteins (Davies, 2004), intracellular thiols (*e.g.* glutathione) (Di Mascio et al., 1991), DNA bases (Cadet et al., 2006) and secondary metabolites. Being one of the strongest oxidizing agents, the radical species •OH is by far the most toxic type of ROS in cyanobacteria. The half–life of •OH is as short as less than 1 μs due to its ability to participate in addition, hydrogen abstraction and electron transfer reactions (Borg et al., 1978). Its high reactivity causes a rather non–selective attack on biological molecules, along with a very short diffusion path from the site of production (Czapski, 1984). The less reactive superoxide anion ($\bullet O_2^-$) is occasionally generated from one electron reduction of O_2 at the receptor–side of electron transport chain during respiration and photosynthesis. Because it is negatively charged, $\bullet O_2^-$ cannot diffuse across the lipid bilayer of biological membranes, but it can pass through certain anion channels. However, $\bullet O_2^-$ can be converted to H_2O_2 by SOD and can release Fe^{2+} from $[4Fe–4S]^{2+}$. H_2O_2 is a relatively stable molecule, but it can produce the highly toxic •OH can then be produced via the $\bullet O_2^-$ promoted Fenton reaction ($Fe^{2+} + H_2O_2 \rightarrow Fe^{3+} + OH^- + \bullet OH$).

Under stress conditions, such as iron starvation, CO_2 limitation, dehydration and UV radiation, the production of ROS in cyanobacteria can be accelerated. However, as far as 2.5–3.5 billion years ago, cyanobacteria were able to manage the oxidative stress, when much higher levels of UV–C and UV–B irradiation could reach the surface of the Earth due to lack of O_2/O_3 protection. Even under today's atmosphere, it has been reported that cyanobacteria including *Chroococcidiopsis* sp. (Cockell et al., 2005; Grilli Caiola and Billi, 2007; Billi, 2009), *Deinococcus radiodurans* (Mattimore and Battista, 1996) and *Nostoc commune* (Shirkey et al., 2003) can survive intense UV radiation and prolonged dehydration, and thrive in extreme habitats such as the hot Negev Desert (Israel) (Büdel and Wessels, 1991) or the cold deserts of McMurdo Dry Valleys (Antarctica), Ross Desert (Antarctica) (Friedmann and Ocampo–Friedmann, 1985) and Atacama Desert (Chile) (Warren–Rhodes et al., 2006; Wierzchos et al., 2006). Other hostile environments are hot springs that are characterized by the absence/scarcity of nearby vegetation, high temperature and the possible presence of toxic compounds (*e.g.* arsenite and arsenate); but also the intertidal marine and hypersaline zones, clear benthic waters with abundant sunlight, the marine and freshwater interfaces, and exposed terrestrial habitats; yet still cyanobacteria can be found living in these environments.

The knowledge of the mechanisms that cyanobacteria adopt to survive the oxidative stress in various environments will not only increase our understanding of photobiology but may also be applied in areas as remote as astrobiology, since ROS will be encountered by phototrophs in space (Grilli Caiola and Billi, 2007). In this chapter, a wide variety of defence strategies for cyanobacteria against ROS has been reviewed. The effects of ROS formation on aquatic ecosystems and drinking water quality are discussed. Special focus is placed on the employment of ROS to alleviate the unwanted effects of cyanobacterial growth on potable water reservoirs and on sheltered historical monuments (Hsieh et al., 2013; 2014).

2. Formation of ROS

Among the various types of ROS, normally only three are taken into consideration in biological environments: superoxide ($\bullet O_2^-$), hydrogen peroxide (H_2O_2) and the hydroxyl radical ($\bullet OH$). However, in photosynthetic organisms such as cyanobacteria, a fourth species of ROS known as singlet oxygen ($^1O_2$) also becomes important (actually there are two types of singlet oxygen, but in the present context they may be considered as one). ROS in cyanobacteria can be generated from visible light illumination, UV rays and dehydration. They can be enhanced by environmental stress such as CO_2 limitation, moderate heat and high salinity.

In photosynthesis, the superoxide anion ($\bullet O_2^-$) is produced as the reduced product of molecular oxygen during the transport of electrons to the acceptor side in Photosystem I (PSI). $\bullet O_2^-$ can be reduced to H_2O_2 by superoxide dismutase (SOD), and the highly reactive $\bullet OH$ is then generated through the Fenton reaction involving H_2O_2 and Fe^{2+}: $H_2O_2 + Fe^{2+} \rightarrow OH^- + FeO^{2+} + H^+ \rightarrow Fe^{3+} + OH^- + \bullet OH$. Singlet oxygen ($^1O_2$) can be generated in the light–harvesting complexes in PSII when the excitation energy is transferred from the excited chlorophylls to molecular oxygen. The other pathway of $^1O_2$ formation originates from photodamaged PSII. In the past 50 years, different mechanisms of photodamage to PSII have been considered, including the acceptor–side hypothesis (Vass et al., 1992), low–light hypothesis (recombination mechanism) (Keren et al., 1997), donor–side hypothesis (Callahan et al., 1986; Eckert et al., 1991; Anderson et al., 1998), mechanisms based on $^1O_2$ formed independently of PSII electron transfer (singlet oxygen mechanisms) (Jung and Kim, 1990; Santabarbara et al., 2002), and most recently the manganese mechanisms proposed in 2005

(Hakala et al., 2005; Ohnishi et al., 2005). In this latter model, the oxygen–evolving complex (OEC) is inactivated primarily by light, especially blue light and UV rays, with subsequent release of Mn ions from the complex. The reaction center of PSII is then inactivated by light absorbed by the photosynthetic pigments. Dysfunction of OEC might allow the free access of molecular oxygen to the reaction–center chlorophyll P_{680}^{+} that cannot be reduced in the normal way by the Mn–depleted OEC; hence the subsequent formation of $^{1}O_2$ may occur (Hakala et al., 2005; Nishiyama et al., 2006).

Cyanobacteria have abundant photosynthetic pigments, in particular chlorophyll *a* (Chl *a*) and phycobiliproteins (PBPs) that consist of phycoerythrin (PE), phycocyanin (PC) and allophycocyanin (APC) (Figure 1). PBPs are light harvesting proteins with covalently bound bilins (tetrapyrrole chromophores) that compensate the limited absorption spectra of Chl *a* (blue light 410–440 nm and red light 680–700 nm). In the PBP light harvesting antennae, PE absorbs and is excited best by green light 510–565 nm; PC by red light 610–630 nm, and APC by red light 650 nm.

These photosynthetic pigments—Chl *a*, PE, PC and APC—with their highly conjugated double bonds can function as natural photosensitizers when they are exposed to the proper wavelengths. The electrons in the chromophores of PBPs can be excited by both visible light (Jun and Kim 1990; Rinalducci et al., 2004; Rinalducci et al., 2008; He et al., 1997; Paul et al., 2006; Zhan et al., 1999a,b) and UV–A (Castenholz and Garcia–Pichel, 2012), and to less extent by UV–B (He and Häder 2002a,b; Rinalducci et al., 2006) to their high–energy state. These excited pigments can exchange electrons or H–atoms with adjacent molecules in the cells, producing radical species. The radicals may further react with molecular oxygen to form $\bullet O_2^-$, H_2O_2 and •OH (Type I reaction). Alternatively, the excited pigments may directly transfer energy to molecular oxygen, producing $^{1}O_2$ (Type II reaction) (Dolmans et al., 2003) (Figure 2).

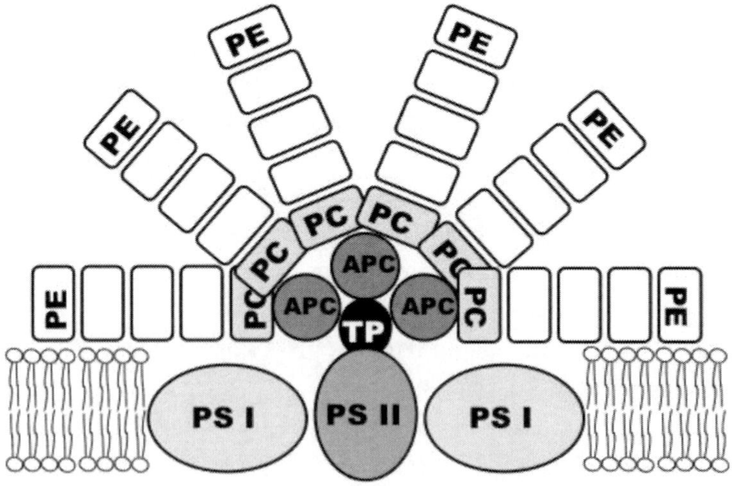

Figure 1. The phycobilisome consists of phycoerythrin (PE), phycocyanin (PC) and allophycocyanin (APC) and embeds a terminal chlorophyll *a* pigment (TP) that connects to PSII.

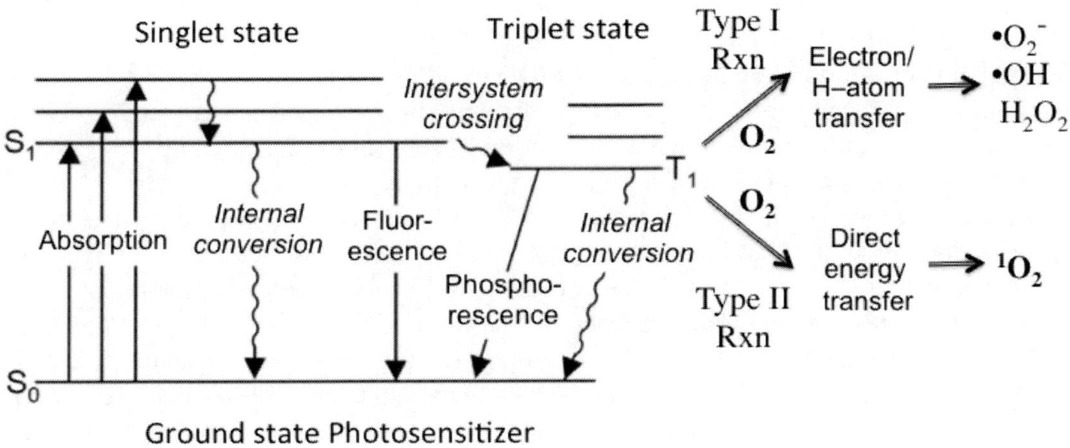

Figure 2. Energy transfer during photosensitization of pigments that generate different ROS via the transfer of electrons/H–atoms to molecular oxygen (Type I reaction) or via direct energy transfer to the molecular oxygen (Type II reaction).

While UV–A (315–400 nm) can sensitize the photosynthetic pigments to produce radicals and ROS in cyanobacteria, UV–B (280–315 nm) mainly produces ROS indirectly by damaging the photosynthetic and respiratory enzymes involved in CO_2 fixation. UV–C (100–280 nm) is absorbed by the ozone layer and scattered in the atmosphere before it can reach the earth. The activity of ribulose–1,5–biphosphatase carboxylase–oxygenase (Rubisco) in the Calvin cycle in the cyanobacterium *Anabaena* sp. was shown to be inhibited by UV–B (Kumar et al., 2003). Because Rubisco catalyzes the production of glycerate–3–phosphate (3–PGA) that is then converted to triose–phosphates at the expense of NADPH and ATP, inhibition of Rubisco may reduce the consumption of NADPH and result in a drop in the $NADP^+$ level (Figure 3). Since $NADP^+$ is the major acceptor of electrons in PSI, depletion of $NADP^+$ will promote the reduction of O_2 to produce ROS (Asada, 1999) in the forms of $•O_2^-$ and H_2O_2 (Radmer and Ollinger, 1980). The PSII complex is another direct target of UV–B, which alters the electron transport at both PSII donor and acceptor sides, resulting in electron leakage and the formation of ROS as reported for *Synechocystis* PCC 6803 (Vass et al., 1999).

Other types of environmental stress that suppress CO_2 fixation in the Calvin cycle can also create ROS in cyanobacteria. Increasing temperature may inactivate Rubisco activase (Feller et al., 1998) that is essential for Rubisco activity, whereas high salinity of NaCl could inactivate Rubisco directly (Solomon et al., 1994). When the Calvin cycle is suppressed, NADPH and ATP cannot be utilized normally, leading to a decrease in $NADP^+$ level that is the main electron acceptor in PSI. The excessive electrons tend to react with molecular oxygen, thereby producing $•O_2^-$ that can then be converted into H_2O_2 by SOD.

Oxidative stress may originate from dehydration. High amounts of radical species in desiccated cyanobacteria was detected by electron spin resonance spectroscopy (Potts, 1994). Although detailed mechanisms of ROS formation in cyanobacteria due to dehydration are still lacking, a study of higher plant chloroplasts, that biochemically and structurally are similar to cyanobacteria, showed increased levels of $•O_2^-$ accompanied by an increase in both intracellular iron concentration and peroxidized lipids during dehydration (Price et al., 1989). Electron leakage from the impaired electron transport chains in the chloroplast could cause the formation of $\bullet O_2^-$ that can be reduced by SOD to H_2O_2 and then react with an electron to generate •OH through the Fenton reaction catalyzed by iron (Price and Hendry, 1991). Dehydration can further promote the accumulation of ROS in the cells, mainly by reducing the hydration shell of biomolecules, thus altering cytoplasmic and intracellular transport. Moreover, ROS formation is accelerated by the shrinkage of cells with the concomitant concentration of all cellular molecules, followed by an increase in ionic strength and fluctuating pH in the cytoplasm (Senaratna and McKersie, 1986). The enzymes in the antioxidant defence systems of cyanobacteria may thus be affected and cannot function normally to scavenge ROS (Billi and Potts, 2002). The photosynthetic apparatus is heavily affected by the limitation of iron since all the membrane protein complexes that catalyze the light reactions contain Fe in the forms of heme, FeS proteins and clusters such as [4Fe–4S] (Martin and Fitzwater, 1988). It was observed that in iron–starved *Anabaena* sp. PCC 7120 cultures, the amount of ROS increased by 10–fold. The amount of ROS also increased drastically in the iron–starved *Synechocystis* sp. PCC 6803. Oxidative damage including lipid peroxidation was observed in these iron–starved cells. However, ROS levels did not change much in the iron–starved *Escherichia coli* and *Bacillus subtilis*, suggesting that iron–starvation–induced oxidative damage is not conserved in heterotrophic bacteria, and could be a characteristic of photosynthetic organisms (Latifi et al., 2005).

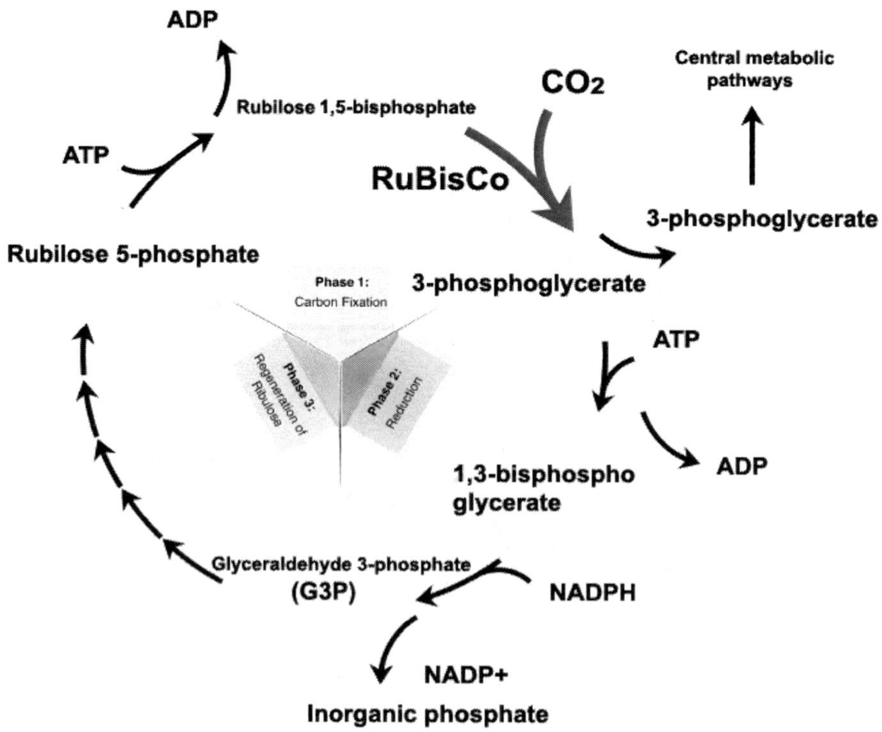

Figure 3. RuBisCo catalyzes carbon fixation in the Calvin cycle. Suppression of the Calvin cycle by environmental stress may affect NADPH and ATP utilization, leading to a drop in the $NADP^+$ level.

3. DEFENCE AGAINST ROS

Cyanobacteria can sustain themselves through oxidative stress conditions by using diverse strategies ranging from ROS prevention and scavenging to reinforcement of photosystems. The formation of ROS can be diminished by the UV–absorbing pigments, by antenna proteins that dissipate excessive light energy as heat, and by migrating away from excessive levels of sunlight. Once ROS are formed, the antioxidant enzymes as well as the nonenzymatic antioxidants can work synergistically as an efficient detoxification factory that eventually convert ROS into H_2O and non–toxic molecules. Their defence mechanisms against ROS can even be strengthened by stabilizing the PSI trimers and chlorophyll–binding proteins, and by promoting the biosynthesis of chlorophyll during PSII turnover/assembly in stressed environment.

3.1. Protective Pigments

Scytonemin: a cyanobacterial sunscreen for near–UV light (300–400 nm)

Scytonemin is a yellowish brown lipid–soluble and non–fluorescent dimeric pigment with a molecular mass of 554 Da, which is specific to cyanobacteria (Figure 4). It is excreted and deposited in the extracellular polysaccharide sheath in more than 300 cyanobacteria species, particularly among the colony-forming populations that get used to high irradiance exposure, such as intertidal mats, epilithic biofilms and biological soil crusts (Garcia–Pichel and Castenholz, 1991). Scytonemin may account for up to 5% of the total dry biomass in cultivated isolates, and the content may be even higher in natural habitats (Sinha et al., 2001). Its complex ring structure with highly conjugated double bonds allows strong absorption of UV–A and blue light, with a maximum around 370 nm *in vivo* (Proteau et al., 1993). The high energy of near–UV is emitted as heat through thermal de-excitation processes by the electrons in the rings, thus shielding the cells from harmful UV radiation from the outer membrane. It is estimated that in the surface filaments of cyanobacteria, the scytonemin-rich sheaths can screen out over 95% UV–A as well as a considerable portion of UV–B (Castenholz and Garcia-Pichel 2012).

Mycosporines and mycosporine–like amino acids (MAAs): pigments for UV–B screening

MAAs (mycosporines and mycosporine–like amino acids) are small, water soluble, colorless compounds composed of a core (either a cyclohexenone or cyclohexenimine) that carries substituents of nitrogen, imino alcohol, or amino acids (Figure 5). MAAs have maximal absorption between 309–362 nm, and can absorb UV–B through electronic transitions in the conjugated enone or enamine portion of the cyclohexene core, which can be modified by the adjacent substituents. The excessive UV radiation is absorbed and dissipated as heat by the conjugated electrons without generating ROS (Conde et al., 2004). In addition to the sunscreen effects, it was reported that mycosporine glycine could quench $^1O_2$ produced by sensitization of Eosin Y based on the clear zone tests on agar plate (Suh et al., 2003). However, whether mycosporine glycine can quench ROS formation inside the living cells still remains unanswered.

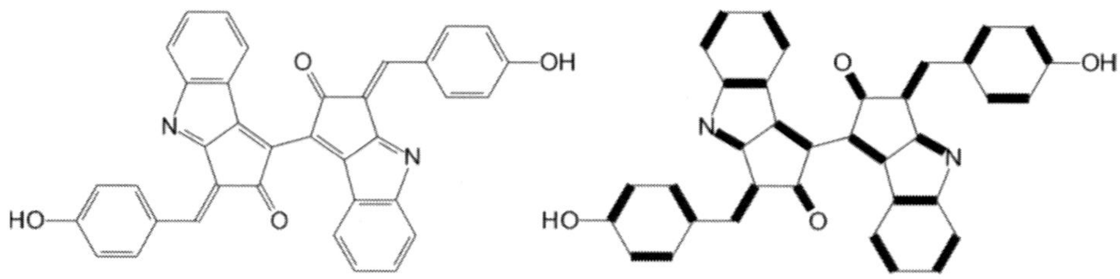

Figure 4. The molecular structure of scytonemin (left). Upon absorption of UV light the molecule does not break down or lose an electron since the energy of the excited state is lowered through the extensive system of conjugated double bonds (right).

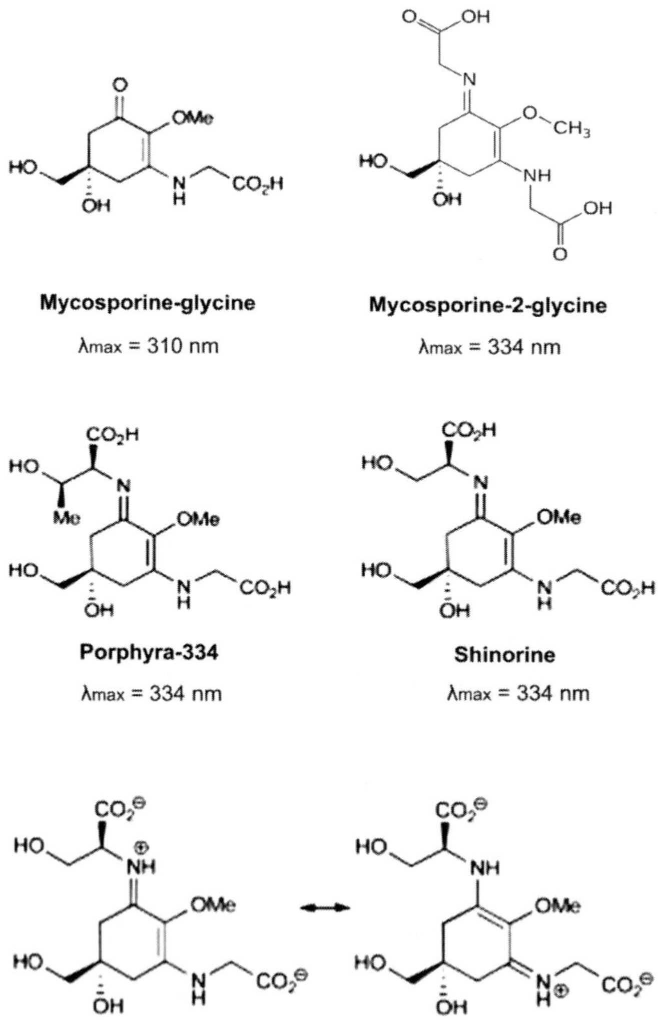

Figure 5. Chemical structures of some representative MAAs in cyanobacteria. Rapid valence tautomerization in the conjugated double bonds reduces the energy of UV–radiation absorbed (bottom).

The protection efficacy provided by sunscreen is closely related to the cellular radius of cyanobacteria. Larger cells offer longer a light–path that contributes to higher absorbance of

UV, according to the Lambert–Beer law. It is estimated that sunscreens provide considerable protection against UV for microplankters (cell radius from 10 to over 100 μm), but could not be the relevant defence mechanism for picoplankters (cell radius less than 2 μm) (Garcia–Pichel, 1994). In addition to scytonemin, other compounds might also function as UV screen in cyanobacteria. The scytonemin–deficient mutant cyanobacterium *Nostoc punctiforme* ATCC 29133 was observed to have similar growth rates as the wild type when exposed to UV–A radiation. Since this strain had no orthologues of the scytonemin biosynthetic genes, this result suggests that some other compound might serve as UV–A protective pigments (Soule et al., 2007). On the cell walls of freshwater green microalgae, for instance, there appear to be biopolymer macromolecules, possibly sporopollenin–type compounds (Castenholz and Garcia–Pichel, 2012), which are typically composed of variable amounts of isoprenoid chains and polyaromatic units fused covalently in a strongly resistant lignin–like matrix. The biological role of such sporopollenin–polymers is unknown, but it seems likely that they too act as sunscreens (Dionisio–Sese, 2010).

3.2. Energy Dissipation

Throughout the history of evolution, phototrophic organisms have developed strategies to survive the oxidative damage induced by ROS during photosynthesis. Cyanobacteria use antenna–related non–photochemical quenching (NPQ) mechanisms to reduce the flux of energy funneled to the reaction center II. The mechanisms include orange carotenoid protein–mediated NPQ, and light energy dissipation mediated by the empty complexes of the chlorophyll–protein IsiA.

Orange carotenoid protein (OCP)

OCP is a 35–kDa water-soluble protein that covalently binds a single carotenoid, 3'–hydroxyechinenone (Kerfeld et al., 2003; Wilson et al., 2010). It was identified as the gene product of the *slr1963* open reading frame in the *Synechocystis* sp. PCC 6803 genome (Holt and Krogmann, 1981; Wu and Krogmann, 1997). Cyanobacteria from diverse habitats over a wide phylogenetic range are found containing homologs of the *Synechocystis slr1963* gene encoding OCP. The protective mechanisms of OCP–phycobilisome in PSII are characterized by the blue–green light induced fluorescence quenching. When cyanobacteria are exposed to high intensity of blue–green light (or white light), the inactivated OCP orange dimer may undergo monomerization into the activated OCP red monomer, resulting in the breakage of the salt bridge and the exposure of the N–terminal domain surface that contains the positively charged Arg155 (Kirilovsky and Kerfeld, 2013; Zhang et al., 2014). OCP is thus bound to the negatively charged sites close to one of the bilin chromophores in the phycobilisome. The excessive light energy can therefore be transferred from the phycobilisome to OCP, from which light energy is dissipated as heat to the environment. Hence, less energy can reach the PSII reaction center, and less light comes back as fluorescence; thus photo–induced oxidative damage is avoided (Kirilovsky, 2007; Wilson et al., 2007; Karapetyan, 2008). After the quenching process, the fluorescence recovery protein (FRP) encoded by the homologs of the *Synechocystis slr1963* gene reverses OPC to its inactivated orange form, restores its tertiary structure and dissociates it from the phycobilisome binding site (Kirilovsky and Kerfeld, 2013; Leverenz et al., 2014) (Figure 6).

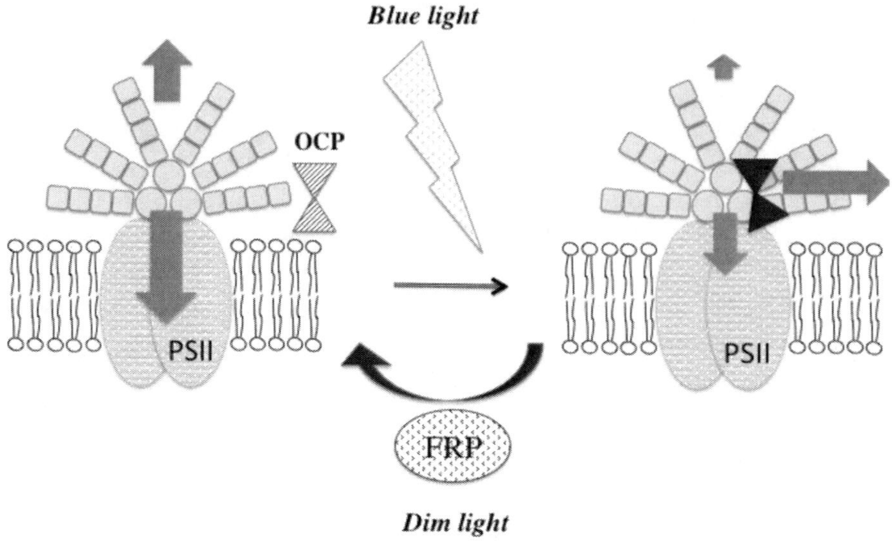

Figure 6. The orange carotenoid protein (OCP) can be induced by blue–green light to change its conformation and bind to the phycobilisome surface. Excessive light energy can be transferred from the phycobilisome to OCP, from which light energy is dissipated as heat, hence avoiding photo–induced oxidative damage to the PSII center. The function of OCP can be monitored through the reduction of autofluorescence emitted from the PSII center. The fluorescence recovery protein (FRP) detaches the OCP from the phycobilisome and recovers OCP to the inactive state in the dark or dim light conditions.

OCP genes are highly conserved among the cyanobacteria possessing phycobilisomes; it was reported that out of 28 cyanobacteria tested, 21 have the entire OCP gene (Boulay and Abasova, 2008). An intact OCP gene is required to perform the blue–green light induced photoprotective mechanism under both iron–repleted and iron–depleted conditions. Strains containing only N–terminal or C–terminal OCP–like genes are sensitive to high light intensities (Boulay and Abasova, 2008). In agreement with this observation, mutants with their OCP gene knocked out became very sensitive to light (Wilson et al., 2006; Wilson et al., 2010). OCP and phycobilisomes are both needed to perform efficient energy dissipation; blue light is unable to quench the fluorescence in the absence of either OCP or phycobilisomes. However, energy dissipation and fluorescence quenching can be induced in the absence of reaction center II or the chlorophyll antennae proteins CP43 and CP47. Therefore, OCP plays a dual role as both the photoreceptor and the mediator for reduction of the energy transferred from phycobilisomes to the PSII center (Wilson et al., 2006).

Iron stress induced antenna protein (IsiA)

Cyanobacteria respond to Fe deficiency and oxidative stress by de–repressing the *isiAB* operon that encodes the antenna protein IsiA and flavodoxin (IsiB). In iron–depleted conditions, a PSI–IsiA supercomplex forms, which is comprised of the PSI trimer encircled by two complete IsiA rings, with 18 IsiA subunits in the inner ring and 25 subunits in the outer ring (Bibby et al., 2001; Boekema et al., 2001; Chauhan et al., 2011) (Figure 7). The ring structure of the IsiA–chlorophyll–binding protein has been found in the peripheries of both PSI and PSII in the cyanobacterium *Prochlorococcus* sp. (Bibby et al., 2003). In addition to iron deficiency stress, it was reported that excessive light (Havaux et al., 2005), high ionic strength (Vinnemeier et al., 1998) and heat stress (Kojima et al., 2006) were able to induce the expression of the *isiAB* operon encoding IsiA protein that protects cyanobacteria from oxidative stress. IsiA has diverse functions including an auxiliary antenna system for PSI (Ryan–Keogh and Macey, 2012), storage of pigment, promotion of the cyclic electron flow around PSI (Ivanov et al., 2000) and protection of cyanobacteria from photodamage by dissipating excessive light energy as heat (Sandström et al., 2001). Although the mechanism(s) of energy dissipation is not yet clear, it is a process different from the blue–green light induced non–photochemical quenching mediated by OCP–phycobilisome (Wilson et al., 2007). By forming the 2 ring–like structures on the PSI trimer, the diameter of the PSI reaction center is increased by around 60% in *Synechocystis* sp. PCC 6803 (Ryan–Keogh and Macey, 2012). IsiA is required to form the gigantic PSI supercomplexes, including the newly identified IsiA–PSI–PSII supercomplex (F4) and IsiA–PSI supercomplex (F5), immensely enhancing the efficency of state transition (Wang et al., 2010), excitation trapping and electron transfer (Ivanov et al., 2000; Chauhan et al., 2011), thereby adapting cyanobacteria to a variety of adverse environments.

3.3. Stabilization of Photosystems: The Role of HLIPs and Small CAB–Like Proteins (Scp)

High light inducible polypeptides/proteins (HLIPs) and small CAB–like proteins (Scp) are encoded by *hli* genes. The terminology of HLIPs and Scp can be used to refer to the same *hli* gene product. The five small CAB–like proteins (ScpA, ScpB, ScpC, ScpD and ScpE) correspond to the five HLIPs (HemH, HliC, HliA, HliB and HliD), respectively (Funk and Vermaas, 1999).

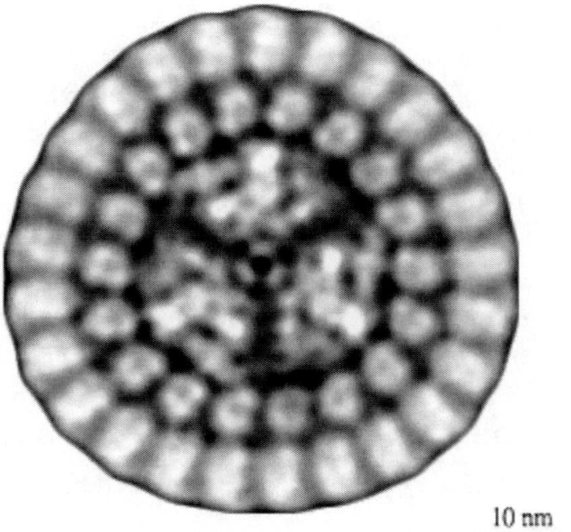

Figure 7. Single–particle TEM image of a gigantic PSI–IsiA supercomplex from *Thermosynechococcus elongatus* grown at nanomolar Fe levels. The outer ring is composed of 25 IsiA subunits and the inner ring of 18 subunits. The PSI trimer is surrounded by these 2 complete IsiA rings. (Source: courtesy of Chauhan et al. 2011. *Biochemistry 50*, 686–692).

HLIPs are critical for cyanobacteria to survive the high light conditions. It was suggested that HLIPs stabilize PSI trimers, interact with Slr1128 (with which HliA and HliB interact), and protect cells from high–light induced oxidative stress (He et al., 2001; Wang et al., 2008). In *Synechocystis* sp. PCC 6803, the PSI activity of the *hli* mutant was reported to be 3 to 4–fold lower than the wild–type strains. The *hli* mutants lost 30–50% PSI trimers after 12 h incubation under the medium–high light, and lost all the PSI trimers after 12 h exposure to high light. The reduction of 77 K chlorophyll fluorescence was also observed in the *hli* mutant strains (Wang et al., 2008).

The small CAB–like proteins (Scp) might play an important role in survival from the oxidative stress. Scp can act as pigment carriers (Yao et al., 2007) and are required in the stabilization of chlorophyll–binding proteins (Vavilin et al., 2007) and in the tetrapyrrole metabolic pathway (Sobotka et al., 2008). It is likely that chlorophyll released from the damaged PSII complex is temporarily stored by Scp during the repair process (Nixon et al., 2010). Scp could also be involved in chlorophyll biosynthesis during PSII turnover/assembly that is essential to ameliorate oxidative stress (Hernández–Prieto et al., 2011).

3.4. Antioxidant Enzymes

Just like almost all other types of cells living under aerobic conditions, cyanobacteria has an intracellular defence against the different ROS species, based mainly on three antioxidant enzymes and a few types of small organic molecules. Essentially three types of antioxidant enzymes, namely superoxide dismutases (SODs), catalases and peroxidases, together with low molecular weight antioxidants, such as ascorbate (Vitamin C), α–tocopherol (Vitamin E) and glutathione (GSH), make up the standard defence system against ROS. Some indirect defence mechanisms are also found, which serve to avoid the generation of ROS; these are typically metal–sequestering proteins, which prevent the single electron reactions of spurious transitions metals, as well as quinone reductases able to perform reductions with 2 electrons in order to avoid the formation of semiquinones and similar radicals.

Any radical molecule has a single unpaired electron, and the reaction of a radical with a normal non–radical molecule will therefore always give rise to the formation of a new radical with an unpaired electron. The single unpaired electron of superoxide can only be eliminated through the reduction of a suitable transition metal, such as the reduction of Cu(II) to Cu(I), or by the reaction with another radical. The first possibility is used in the superoxide dismutases that normally use either Cu, Fe or Mn in the active site, and react with two superoxide anions in two consecutive steps:

$$\bullet O_2^- + SOD(II) \rightarrow O_2 + SOD(I) \text{ (oxidation of superoxide)}$$

$$\bullet O_2^- + SOD(I) + 2H^+ \rightarrow H_2O_2 + SOD(II) \text{ (reduction of superoxide)}$$

Together these two half–reactions produce the dismutation of two superoxide ions, one is oxidized to molecular oxygen and the other is reduced to hydrogen peroxide. Many different SODs have been identified in cyanobacteria (Priya et al., 2007; 2010), mainly Fe– and Mn–enzymes but also a few Cu/Zn–SODs. In addition 17 putative Ni–SODs have been identified from their protein sequences, although it is still not clear whether superoxide dismutation is the true physiological role of these enzymes. The importance of SODs in cyanobacteria is shown by the fact that various types of stress have been found to induce marked increases in SOD contents (Kim and Suh, 2005; Priya et al., 2010; Shirkey et al., 2000). Consistently, strains lacking SOD are much more susceptible to oxidative stress (Thomas et al., 1998). The mechanisms behind appear to be complicated; in *Microcystis aeruginosa* it was found that Fe–SOD increased due to photooxidative stress but Mn–SOD did not, and the high concentrations of ROS generated under prolonged photooxidative stress were able to inactivate and degrade the Fe–SOD (Canini et al., 2001).

Hydrogen peroxide is not a radical species, and is eliminated by catalase in another dismutation process somewhat similar to the one used by SODs, but the mechanism is more complicated here since there are two electrons that need to be transferred from one H_2O_2 to another H_2O_2. Also in this case the oxidized product is molecular oxygen, but the reduced product is now water. The details of the overall process is as follows:

$$H_2O_2 + \text{catalase(IV)} \rightarrow O_2 + \text{catalase(II)} + 2H^+ \text{ (oxidation of hydrogen peroxide)}$$
$$H_2O_2 + \text{catalase(II)} + 2H^+ \rightarrow 2H_2O + \text{catalase(IV)} \text{ (reduction of hydrogen peroxide)}$$

SODs and catalase are among the most efficient enzymes known due to their high reactivity combined with the short diffusion times of the substrates, but also because of their ability to react first with one substrate molecule and then wait for the second molecule to arrive. This results in an efficient scavenging of $\bullet O_2^-$ and H_2O_2 inside the cyanobacteria, where the concentrations of SOD and catalase are likely to be several orders of magnitude higher than the $\bullet O_2^-$ and H_2O_2 levels. In addition to catalases, cyanobacteria may also use different types of peroxidases to eliminate peroxides, such as glutathione peroxidases, ascorbate peroxidases and peroxiredoxins (Hosoya–Matsuda et al., 2005; Tel–Or et al., 1986; Dietz, 2010). The level of these enzymes has also been reported to increase under stress conditions (Chaneva et al., 2009), and disruption of the genes drastically lowered the viability of the cyanobacteria even under low light conditions (Hosoya–Matsuda et al., 2005).

The main damage caused by ROS, however, is believed to come from •OH and R–O• (hydroxyl and alkoxyl radicals) formed when spurious transition metal ions reduce hydrogen peroxide or other peroxides:

$$H_2O_2 + Fe^{2+} \rightarrow OH^- + \bullet OH + Fe^{3+}$$
$$R\text{–}OOH + Fe^{2+} \rightarrow OH^- + R\text{–}O\bullet + Fe^{3+}$$

The first of these equations is known as the Fenton reaction, and in cyanobacteria such Fenton–type reactions are likely to be carried out by Fe(II) and Cu(I). In theory, also other redox–active metals can be Fenton–active, but normally they will be present in too low amounts to have importance or they will have too unfavorable oxidation potentials to be active.

•OH in particular is extremely reactive and in practice it may react with any kind of biomolecule that it encounters; the targets can be proteins, lipids, carbohydrates or nucleic acids. Typically, the reaction with •OH will either insert a covalent hydroxyl group on the target molecule or abstract a hydrogen atom to form H_2O, but it always forms another very reactive radical species. Neither cyanobacteria nor other cells have any specific antioxidant defence against the hydroxyl radical, perhaps with the exception of GSH, which is likely to be the target because it reaches millimolar concentrations in cyanobacteria (Tel–Or et al., 1985).

Due to their photosynthetic activity, cyanobacteria are also at risk of producing singlet oxygen ($^1O_2$), as described in the previous section (2. Formation of ROS). $^1O_2$ is another very reactive type of ROS against which there is no specific protection available. Therefore, cyanobacteria have to face two of the most reactive and dangerous ROS species, $^1O_2$ and •OH, without having any suitable antioxidant defence mechanisms. In practice, the only defence is to avoid the formation of these ROS species in the first place.

3.5. Nonenzymatic Antioxidants

In addition to the antioxidant enzymes there are also small organic molecules with antioxidant activity in cyanobacteria. Some of them are the standard antioxidants found almost ubiquitously in cells, such as ascorbate and GSH in the cytosol, and the various tocopherols (together known as Vitamin E) found exclusively in the membrane lipid bilayer, together with rare types of other non-structural membrane lipid components. Some cyanobacteria produce different small hydroxylated aromatic compounds that typically are species–specific, and which may show antioxidant activity when tested *in vitro;* however, these molecules normally also are toxins, and at present there is no evidence that they may play any significant role as antioxidants *in vivo*. But even classic antioxidant scavengers like ascorbate are unlikely to protect cells against ROS. The level of ascorbate in cyanobacteria is far too low (20–100 μM; Tel–Or et al., 1985) to make it a useful scavenger of •OH, and H_2O_2 is not a radical and is not affected by ascorbate. SOD is a much more efficient intracellular scavenger of superoxide than ascorbate; $•O_2^-$ might be reduced by ascorbate, but this would lead to one H_2O_2 formed from each superoxide anion with a 1:1 stoichiometry, whereas SOD produces a more favorable 0.5:1 ratio because half of the superoxide is oxidized to O_2. This is a general problem with all nonenzymatic reducing antioxidants: there is not much advantage in transforming $•O_2^-$ to the potentially much more problematic H_2O_2, from which the hydroxyl radical can be generated.

The reaction of GSH with $•O_2^-$ is even more unfavorable, because the superoxide is not reduced but regenerated in a chain reaction that transforms reduced glutathione to its oxidized dimer form, GSSG (Winterbourn and Metodiewa, 1994). In practice this has little importance, since also in this case the rate constant of the reaction is far too low to allow competition with SOD, whenever the active enzyme is present (Jones et al., 2003). However, GSH and its thiol group may actually have an important role as a possible scavenger of the hydroxyl radical and singlet oxygen in cyanobacteria, because it is one of the few cell components present at millimolar concentrations (Tel–Or et al., 1985). It is therefore a likely target for the spontaneous reactions of $^1O_2$ and •OH, and in this sense GSH may protect against ROS, although its antioxidant activity here is completely passive. The real antioxidant effect of ascorbate and GSH is likely to be their function as the substrates that provide ascorbate peroxidase and GSH peroxidase with reducing power.

Carotenoids are tetraterpenoids possessing 40 carbon atoms and 3–13 (usually 9–13) conjugated double bonds. More than 750 carotenoids have been structurally identified in nature (Britton et al., 2008). The most common carotenoids in cyanobacteria include β–carotene, zeaxanthin, echinenone and myxolpentosides (myxoxanthophyll) (Takaichi and Mochimaru, 2007). They function as antioxidants by quenching both triplet excited chlorophylls (^{3}Chl*) and $^1O_2$ (Triantaphylidès and Havaux, 2009), scavenging free radicals, and suppressing lipid peroxidation (Johnson et al., 2007). Carotenoids are excellent energy sinks and dissipaters that accept excess light energy from excited chlorophylls, IsiA–PSI supercomplex and OCP–phycobilisome structure due to their sufficiently low–lying singlet–excited state (S_1). On a timescale of a few picoseconds, the S_1 carotenoids can then dissipate the excess energy by decaying to the ground state (S_0) (Berera et al., 2009). Similarly, carotenoids can deactivate the excited chlorophylls via a charge separation mechanism. Excess energy is transferred from excited chlorophyll molecules to a chlorophyll–zeaxanthin heterodimer, which then undergoes charge separation to dissipate energy. The following scheme of energy flow has been proposed:

$$\text{Chl}^*_{\text{pool}} \rightarrow (\text{Chl–Zea})^* \rightarrow \bullet\text{Chl}^- + \bullet\text{Zea}^+ \rightarrow \text{Chl–Zea}$$

Deactivation of singlet excited state chlorophyll (^{1}Chl*) under excess light occurs when the excitation energy is transferred to a chlorophyll–zeaxanthin heterodimer (Chl–Zea), followed by an ultrafast formation of zeaxanthin cation (•Zea$^+$) (Holt et al., 2005).

3.6. Vertical Movement

Gliding filamentous cyanobacteria as well as some unicellular species in soft microbial mats and sediments can move towards/against sunlight irradiance, so that the optimal photosynthetic rate with minimal photooxidative damage can be approached. For instance, the filamentous cyanobacteria *Oscillatoria* sp. and *Spirulina subsalsa* in hypersaline benthic mats have been demonstrated to migrate downwards to the lower layers of the mat in the daytime and rise to the surface layer at dusk (Garcia–Pichel et al., 1994). *Phormidium uncinatum* was observed to migrate away from the damaging irradiation *in vitro* (Donkor et al., 1993). Most of these cyanobacteria investigated contain little or no MAAs and cannot withstand large doses of UV. The vertical movements might be regulated by UV and blue light exposure, especially UV–A (Kruschel et al., 1998). The shape and length of cyanobacterial trichomes contribute to their gliding motility, allowing them to escape from photo–induced damage (Tamulonis et al., 2011).

4. Physiological Impact

ROS in cyanobacteria may cause photoinhibition by hindering the repair mechanism of the photodamaged PSII. The same wavelengths of light used for photosynthesis may produce ROS by sensitizing chlorophyll *a* and the bilin chromophores in phycobiliproteins, resulting in ROS attack and photobleaching of pigments. DNA breakage is observed in desiccated or UV–irradiated cyanobacteria accumulating a high level of ROS. Thylakoid peroxidation occurs when the unsaturated fatty acids in the thylakoid membrane are broken down by ROS via oxygen–dependent free radical chain reactions. A loss of membrane integrity appears when the rate of cellular damage surpasses the capacity of the repair machinery. In addition to causing oxidative cell lesion, ROS can trigger the genetically regulated programmed cell death in cyanobacteria.

4.1. Photoinhibition

In cyanobacteria and other aerobic cells, the extent of oxidative stress depends on the balance between ROS formation and ROS scavenging. Under normal conditions, the rate of ROS scavenging equals the rate of ROS formation. However, the balance is disturbed when cyanobacteria are faced with an environmental stress, such as high intensity of light, CO_2 limitation, high salinity, excessive heat or low temperatures (Murata et al., 2007; Takahashi and Murata, 2008). When light intensity reaches the saturation point, for example, any further increase in light will lead to a decrease of the photosynthetic rate, a phenomenon known as 'photoinhibition' (Aro et al., 1993).

Although it had been generally accepted at an early stage that photoinhibition of PSII is the result of ROS attack, the action of ROS was actually misunderstood for a long time. It was believed that ROS cause photoinhibition by directly damaging the PSII. However, by separating the reaction of photodamage of PSII from the repair of the photodamaged PSII, it has been clarified that the chief role of ROS is the inhibition of the synthesis *de novo* of D1 protein that is required in the repair process of the damaged PSII. The inhibition regards the elongation step during translation of *psbA* mRNA (Samuelsson et al., 1985; Nishiyama et al., 2006). In contrast, the extent of photodamage in PSII is exactly in proportion to the intensity of light, which is not initiated by ROS (Tyystjärvi, 2008). Photoinhibition occurs when the rate of photodamage of PSII is accelerated by high light and/or the rate of repair of the damaged PSII is suppressed by ROS.

Except for high intensity of light, photoinhibition can be intensified by other kinds of environmental stress, such as CO_2 limitation, high salinity, heat and low temperature, that create additional ROS production by suppressing the Calvin cycle in photosynthesis (Takahashi and Murata, 2008). The elevated levels of ROS generated could hinder the repair mechanism in the damaged PSII, indirectly leading to photoinhibition.

4.2. Photobleaching

Exposure of cyanobacteria to medium intensity UV–B as well as to strong visible light can cause ROS formation from the excited phycobiliproteins and chlorophyll *a*, which eventually are destructed by ROS generated by the pigments themselves. This phenomenon is referred to as 'photobleaching'.

As mentioned above, $^1O_2$ is highly reactive towards the amino acid residues His, Trp, Met and Cys (Halliwell and Gutteridge, 1990). H_2O_2 is generally less reactive; however, it can indirectly damage proteins by generating •OH through the Fenton reaction due to the presence of protein–bound metal ions (Fridovich, 1998). These highly reactive oxygen species are generated via both Type I and Type II photoreactions (Paul et al., 2006). They tend to attack the apoprotein in phycobilisomes, leading to cleavage of the peptide bonds, but also to cross–linking and aggregation reactions. In comparison $\bullet O_2^-$ is far less reactive, and although it can be generated in Type I reactions, it is not believed to be directly involved in the photobleaching process. By using spin trapping ESR spectroscopy, the formation of ROS was detected from the phycobilisomes isolated from *Synechocystis* PCC 6803. The light–harvesting antennae underwent unspecific and random cleavage of biliprotein peptide bonds that were directly visible as a smear in the SDS–PAGE gel (Rinalducci et al., 2006; 2008).

4.3. DNA Breakage

Oxidative stress causes DNA breakage through single electron/H–atom transfer (Type I photoreaction) and singlet oxygen oxidation (Type II photoreaction). The preferential substrates for ROS attack during Type I reaction are the base moieties ranking from guanine, adenine, thymine, cytosine to 5–methylcytosine. A so–called hole transfer was found to occur in double–stranded DNA (Schuster, 2000; Giese, 2000), mainly from pyrimidine and adenine radical cations to guanine, which serves as a sink. Relatively high amounts of guanine radical cations are formed, depending on the base sequence. These base radical cations are subjected to hydration and deprotonation reactions (Ravanat et al., 2001).

In Type II reactions, 2'–deoxyguanosine (dGuo) is the sole known DNA target for oxidation by $^1O_2$ (Cadet and Téoule, 1978), probably involving the formation of transient diastereomeric endoperoxides (Sheu and Foote, 1993). The main product, 8–oxodGuo is formed by rearrangement of a 4,8–endoperoxide intermediate into 8–hydroperoxy–2'–deoxyguanosine, and followed by reduction. 8–OxodGuo can be decomposed, forming cyanuric acid nucleoside as the main product (Raoul and Cadet, 1996).

In addition, ROS could indirectly lead to DNA breakage by activating Ca^{2+}–dependent endonucleases as a result of elevated levels of intracellular free Ca^{2+}, and by interfering with the enzymes involved in DNA replication or repair (Halliwell and Gutteridge, 1999). Severe DNA damage could induce necrosis and apoptotic cell death (Halliwell and Gutteridge, 1999).

4.4. Thylakoid Degradation

Thylakoid degradation is the most evident alteration in the ultrastructure of cyanobacteria subjected to oxidative stress (Schmetterer et al., 1983; Baulina et al., 2004). The primary sites of ROS attack on thylakoid membrane are the unsaturated fatty acids that possess multiple double bonds and active hydrogens. Decomposition of thylakoid membranes is caused by a radical chain reaction with three steps. The initial step (1) begins when the fatty acid is attacked by ROS such as •OH and •OOH to produce a fatty acid radical. A propagation step (2) follows where the fatty acid radical reacts with molecular oxygen to produce a peroxyl–fatty acid radical. This radical reacts further with another fatty acid molecule in the last step (3) to regenerate the fatty acid radical and form a lipid peroxide, which can participate in Fenton-type reactions and produce other fatty acid-derived radicals. This chain reaction can only be terminated through reduction of the radicals by antioxidants like vitamin E, whereas the lipid peroxides must be removed by specific peroxidases (Figure 8).

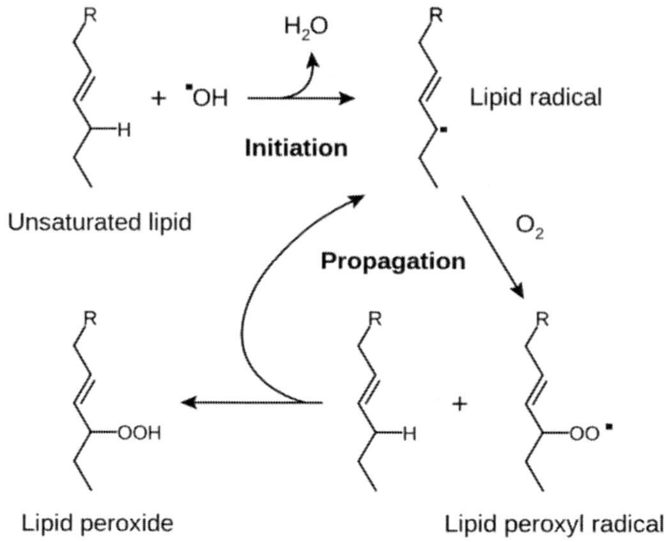

Figure 8. Lipid peroxidation via free radical chain reactions initiated by hydroxyl radical attack. The lipid radicals produced are then oxidized by O_2 into lipid peroxyl radicals, which can react with a new lipid molecule to regenerate a lipid radical, thus propagating the radical chain reactions and resulting in the oxidative damage of the membrane.

The degradation products of thylakoid are high amounts of ethane (Elstner and Pils, 1979; Schmetterer 1982) and malondialdehyde (MDA) (Zeeshan and Prasad, 2009). MDA may cause mutations in cyanobacteria by reacting with deoxyadenosine and deoxyguanosine in DNA to form DNA adducts, mainly M_1G. The guanidine group of arginine residues may condense with MDA to give 2–aminopyrimidines (Marnett, 1999).

4.5. Programmed Cell Death (PCD)

Oxidative stress, particularly H_2O_2 may trigger apoptosis–like programmed cell death in cyanobacteria by causing oxidative damage to DNA or by indirect oxidation of regulatory molecules that ultimately serve as an entrance to PCD (Slater et al., 1995). PCD is an irreversible, genetically controlled form of cell suicide and is categorized as apoptosis, paraptosis, aponecrosis and autophagy. Apoptosis features DNA fragmentation, chromatin condensation, plasma membrane blebbing and an increase in caspase activities. Caspases are a unique family of cysteine aspartate–specific proteases that initiate PCD and subsequently activate other degradative enzymes, including endonucleases that cause DNA fragmentation. Caspases activity is a hallmark of apoptosis and can be measured through the cleavage of DEVD, a caspases–specific substrate. Readers interested in the mechanisms of cell death under oxidative stress are referred to the comprehensive review with 451 citations by Ryter et al. (2007).

Apoptosis–like PCD has been observed in the cyanobacteria *Microcystis aeruginosa* (Ding et al., 2012; Ross et al., 2006; Bouchard and Purdie, 2011; Kaneko et al., 2007; Frangeul et al., 2008), *Trichodesmium* spp. (Berman–Frank et al., 2004) and *Anabaena* sp. (Ning et al., 2002). H_2O_2 may induce apoptotic cell death in a dose–dependent manner.

While 150 μM H_2O_2 had only weak effects on *Microcystis aeruginosa* proliferation, cells treated with 250 and 325 μM H_2O_2 displayed characteristic apoptosis features, such as membrane deformation, partial disintegration of thylakoids, cytoplasmic vacuolation, nucleoid chromatin condensation, elevated caspase–3–like activity and DNA fragmentation, which could be intensified by higher concentration of H_2O_2 (Ding et al., 2012).

PCD plays an important role in the developmental and differentiation processes as part of an adaptation to environmental stress. In cyanobacteria, PCD is needed in the formation of heterocyst cells in diazotrophic species, such as *Anabaena* spp., and in the formation of hormogonia that are released from the parent filaments after death and serve as dispersal and infective units in plant–cyanobacterial symbioses.

5. Environmental Impact and Applications

ROS in cyanobacteria are responsible for coral bleaching in scenarios of global warming and ocean acidification. ROS may induce the seasonal disappearance of algal blooms and eutrophication in water bodies, threatening the ecosystems and human populations by releasing cyanotoxins and off–odor compounds from the decaying cyanobacteria and microalgae. Nonetheless, ROS can be employed in water purification processes and can be regarded as a safe and effective tool to protect historical monuments from cyanobacterial attack. Furthermore, ROS can enhance electricity generation in bio–electrochemical systems while H_2O_2 produced by cyanobacteria under oxidative stress has potential to be harvested as renewable energy from solar radiation.

5.1. Coral Bleaching

The reef–building corals are known to live an endosymbiotic life with microalgae, dinoflagellates (e.g. *Symbiodinium* spp., zooxanthellae) and cyanobacteria (e.g. *Phormidium corium*). The vegetative symbionts may constitute up to 30% of total coral tissues in some reefs. These phototrophic symbionts absorb CO_2 and nitric compounds discharged from the coral host to perform photosynthesis and metabolism; meanwhile providing food and energy to the corals.

However, the photosynthetic nature of the symbionts make them sensitive to environment–induced oxidative stress, such as UV and photosynthetically active radiation (PAR), salinity variation and changes in temperature, particularly the temperature of seawater (Baker 2008). It has been demonstrated that even the most tolerant corals can only sustain temperature increases of 2–3 °C above their long–term solar maxima for short periods (Hoegh–Guldberg, 1999; Loya et al., 2001).

Rapid climate changes along with ocean acidification put a serious threat to the delicate coral–reef ecosystems. The most notable examples of coral bleaching may be the Great Barrier Reef in Australia and the coral reefs along the coast of the Indian Ocean, such as the Maldives, the Andamans, and the Lakshadweep and Seychelles Marine Park System. In the outer–atoll seaward slopes of Lakshadweep Island in India, only 3% coral cover survives (Bhandari and Sharma, 2010).

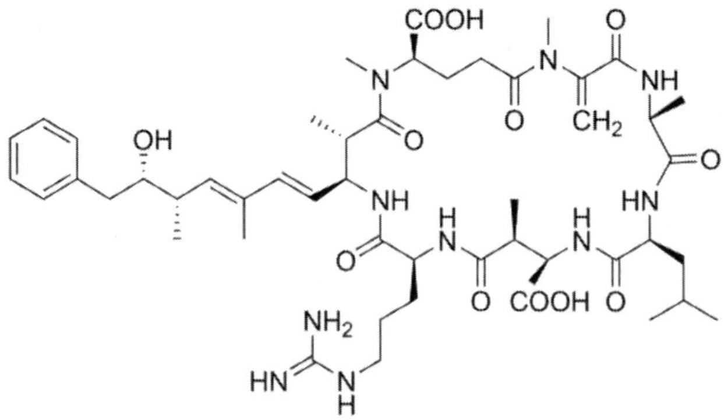

Microcystin–LR

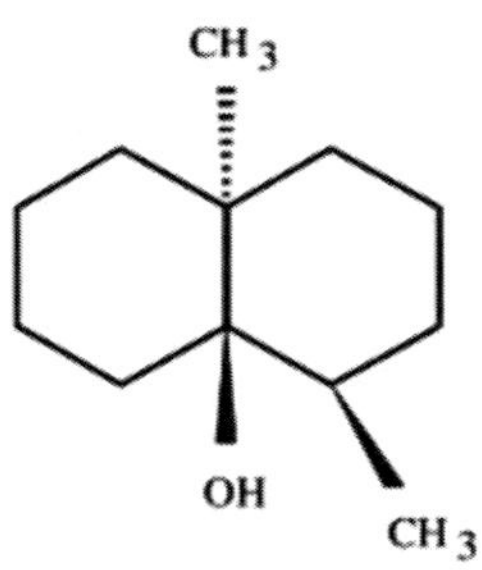

Geosmin

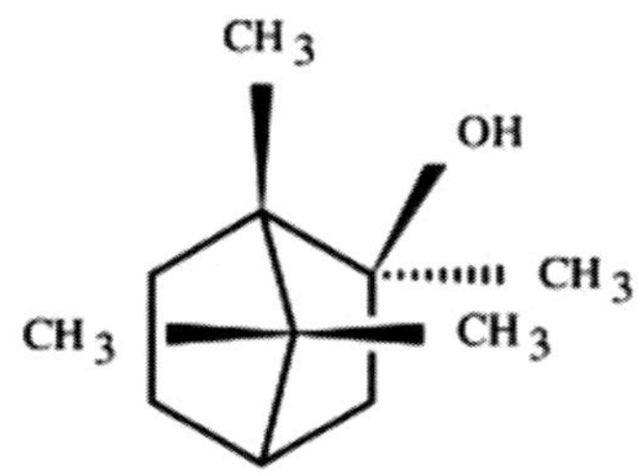

2–methylisoborneol

Figure 9. Chemical structures of microcystin–LR, 2–methylisoborneol and geosmin.

The severe coral–bleaching phenomenon in Lakshadweep Island could be attributed to oxidative stress in the symbiotic cyanobacterium *Phormidium corium*. This strain is reported to be susceptible to ROS attack due to low SOD and ascorbate peroxidase activities, and insufficient levels of ascorbic acid. The amount of UV–B sunscreen MAAs also decline under the combination of UV–B and PAR exposure (Bhandari and Sharma, 2010). Because this cyanobacterial symbiont can no longer grow and provide food for the corals, they will eventually be expelled by the host or might be simply killed by accumulated ROS-induced damage (Takahashi and Murata, 2008). As a result, the bleaching and starvation of corals may occur; leading to the death of coral reefs in the long term.

5.2. Harmful Algal Blooms and Water Purification

Algal blooms reoccur worldwide in the oceans, freshwater bodies and manmade reservoirs as a boom of brown, green, purple, or pink microalgae and cyanobacteria. Harmful algal blooms may deplete oxygen in the water, causing asphyxia in fishes. Harmful algal blooms contaminate the sea–grasses and zooplanktons by releasing hepatotoxins such as microcystins from cyanobacteria, and neurotoxins like brevetoxin from dinoflagellates and domoic acid from diatoms. The death of dolphins and manatees after ingesting the algal bloom–contaminated food has been reported (Flewelling et al., 2005). Sea–turtles also suffered from lethargy and muscle weakness after feeding on the contaminated prey. In addition, the off–odor compounds geosmin and 2–methylisoborneol produced by algal blooms and eutrophication would taint the drinking water with strong earthy–muddy (geosmin) or musty (2–methylisoborneol) odor that raises consumers' complaints (McGuire 1995) (Figure 9).

The seasonal retreat of algal blooms might be due to the ROS–induced programmed cell death. As in the case of the marine cyanobacteria *Trichodesmium* spp., PCD is induced by high irradiance intensities, phosphorus and iron starvation, and oxidative stress that are responsible for the demise of an algal bloom (Berman–Frank et al., 2004). However, it was reported that oxidative stress, cellular injury and PCD in the cyanobacterium *Microcystis aeruginosa* in freshwater could trigger the release of microcystins into the water, rendering it even more toxic for human use (Ross et al., 2006).

Among the detoxification and deodorization methods available, advanced oxidation processes could be the most promising and effective. It is based on the reaction of ROS, especially •OH, with organic compounds by addition reactions or by hydrogen abstraction to form a carbon–centered radical, which may further react with O_2 to form a peroxyl that can undergo subsequent reactions to ultimately produce a variety of oxidation products, including ketones, aldehydes and alcohols (Al–Rasheed, 2005). When a photocatalyst such as TiO_2, ZnO or WO_3 is illuminated with an appropriate wavelength of light, pairs of electrons (e^-) and electron holes (h^+) are generated by photons on the surface of the catalyst. These pairs react with molecular oxygen, water or hydroxyl groups adsorbed on the surface of the catalyst to produce ROS, such as •OH, $\bullet O_2^-$ and, HOO• (Gaya and Abdullah, 2008). The following reactions were proposed for a typical photocatalytic reaction (Figure 10):

$$\text{Catalyst} + h\nu \rightarrow e^- + h^+$$

$$(O_2)_{ads} + e^- \rightarrow \bullet O_2^-$$

$$H_2O \rightarrow OH^- + H^+$$

$$\bullet O_2^- + H^+ \rightarrow HOO\bullet$$

$$HOO\bullet + e^- \rightarrow HO_2^-$$

$$HO_2^- + H^+ \rightarrow H_2O_2$$

$$H_2O + h^+ \rightarrow \bullet OH + H^+$$

$$OH^- + h^+ \rightarrow \bullet OH$$

The off–odor compound geosmin was shown to be degraded by a TiO_2 semiconductor–particle–catalyzed reaction into acyclic alkanones and esters as intermediate products before complete mineralization (Bamuza–Pemu and Chirwa, 2012). The effects of TiO_2 can be enhanced by combining TiO_2 with carbonaceous materials that may inhibit electron–hole recombination and may broaden the light–harvesting ability of the catalyst to the visible light range. For example, the synthesized Graphene oxide TiO_2 composite was recently shown to be efficient in degrading the hepatotoxin microcystin–LR and the off–odor compounds geosmin and 2–methylisoborneol under both UV–A and solar radiation (Fotiou et al., 2013). Another nano–photocatalyst Pt/WO_3 was shown to be able to degrade microcystin–LR under simulated sunlight and to inhibit the growth of *Microcystis aeruginosa* that is responsible for many cases of eutrophication (Zhao et al., 2012). Attempts to control the algal blooms in eutrophic waters have been made using TiO_2–coated hollow glass beads floating in the rivers and exposed to sunlight. By irradiating these TiO_2–coated glass beads with UV–A, the cyanobacteria *Microcystis* sp. and *Anabaena* sp. in a batch reactor model system lost their photosynthetic activity, and their cells separated into individual ones (Kim and Lee, 2005). Photocatalysis has potentiality in controlling the aeroterrestrial microalgal and cyanobacterial biofilms on buildings and other manmade surfaces. For example, •OH from a TiO_2 photocatalyst activated by concurrent irradiation with white fluorescent light and UV light is able to diminish the adhesion of the filamentous cyanobacterium *Phormidium tenue* on glass plates (Ohko et al., 2009). Also ZnO activated with low intensity of UV radiation can produce •OH, effectively interrupting the metabolism and changing the morphology of the microalga *Stichococcus* sp. (Gladis et al., 2010).

5.3. Renewable Resources from Solar Energy

H_2O_2 production

It was reported that among 38 cyanobacteria tested, more than 20 were able to produce H_2O_2 in under red light exposure. Three different kinetics were observed although the reaction sites were not identified (Stevens et al., 1973). Other researchers demonstrated that in the cyanobacterium *Anacystis nidulans*, the production of H_2O_2 was carried out by the photoreduction of O_2 in a Mehler reaction, which was critically affected by the metabolic conditions of the cyanobacteria (Roncel et al., 1989; Mehler 1951a,b). The production yield of H_2O_2 from illumination of cyanobacteria can be raised by the addition of methyl viologen that functions as electron acceptor at or near the terminal acceptors of PSI. The reduced methyl viologen may in turn reduce O_2 to $•O_2^-$, which is then converted by SOD to H_2O_2. To optimize the production of H_2O_2, it is important to remove H_2O_2 from the reactors in real time; otherwise the electron transport chain would be damaged by H_2O_2.

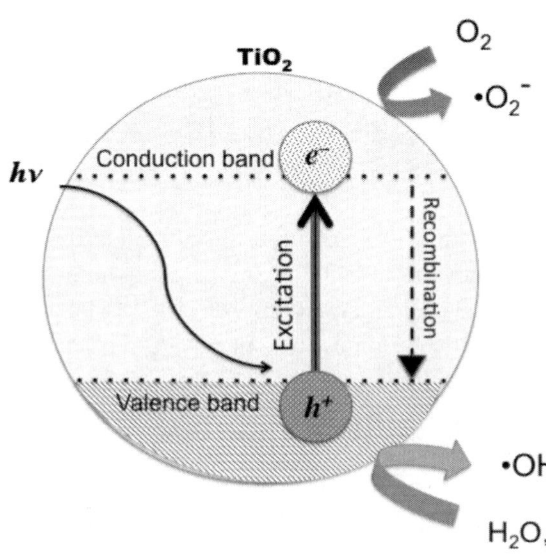

Figure 10. Schematic diagram showing photocatalytic actions of TiO_2 semiconductor particles and the formation of ROS.

Electricity generation with a cyanobacterial biocathode

ROS generated by cyanobacterium *Microcystis aeruginosa* IPP under illumination were able to act as electron acceptor in the *M. aeruginosa* IPP biocathode in bio–electrochemical systems (Cai et al., 2013). Illumination of the *M. aeruginosa* IPP biocathode produced both O_2 and ROS as electron acceptor in photosynthesis. However, it was calculated that when the concentration of ROS in the biocathode is higher than 3.2×10^{-19} M, ROS will become a more efficient electron acceptor than O_2.

The hypothetical cathodic reactions and the standard electrode potentials have been described (Carrette et al., 2000; Campos–Martin et al., 2006):

$$H_2O_2 + 2H^+ + 2e^- \rightarrow 2H_2O,\ \varphi^- = 1.776\ V$$
$$O_2 + 4H^+ + 4e^- \rightarrow 2H_2O,\ \varphi^- = 1.229\ V$$
$$O_2 + 2H^+ + 2e^- \rightarrow H_2O_2,\ \varphi^- = 0.695\ V$$

After dosing the hydroxyl radical scavenger mannitol into bio–electrochemical systems the electrical current dropped sharply, implying an important role for •OH in the generation of electricity in bio–electrochemical systems with a cyanobacterial biocathode (Cai et al., 2013).

5.4. Cultural Heritage Conservation

Cyanobacteria are pioneer organisms capable of colonizing the surface of mineral substrata and growing in the adverse environment of either outdoor monuments exposed to 2000–3000 PAR of sunlight in tropical zones, or in the dim light surroundings such as catacombs, caves and underground chambers. Regarding the ever–increasing concerns of environmental protection, applying biocides on cultural heritage is strictly restricted in the European Union and other countries. Physical methods are instead investigated to safeguard the cultural heritage from biodeterioration problems. Electron holes and ROS created by the excited nano–semiconductors as described in the previous section (5.2 Harmful algal blooms and water purification) have the potential to oxidize the thick, dark brown cyanobacterial biofilms commonly found on outdoor monuments, whereas ROS generated from the excited pigments in cyanobacteria are capable of inhibiting cyanobacterial growth on the covered, dim archaeological sites.

This section discusses the recent advances in the field of cultural heritage conservation that employs visible light to generate ROS in the cyanobacteria colonizing the surfaces of sheltered historical monuments.

Generation of ROS from cellular extracts exposed to light

Cyanobacteria have abundant amount of chlorophyll in thylakoid membrane and abundant phycobiliproteins, including phycoerythrin, phycocyanin, allophycocyanin and phycoerythrocyanin in the light–harvesting antenna system, which can absorb and be excited by a wide spectrum of low intensity of light. Under stressed conditions, such as high light, CO_2 limitation, high salinity and iron starvation, ROS can be generated from these excited pigments in a rate beyond the quenching capacity of the cells, leading to oxidative stress and damage to the cells, as confirmed by electron spin resonance (ESR) spectroscopy (Figure 11).

The results showed that whereas red light can excite pure PC to produce ROS, blue, green and white lights can produce ROS from a mixture of Chl *a*, carotenoids and phycobiliproteins with the possible presence of antioxidants in the crude extracts. Blue or white light was the most efficient to generate ROS in the crude pigment extracts. Although 80 PAR of green light could not generate ROS in the extracts from *Oculatella subterranea*, ROS have been detected in the same system exposed to 150 PAR of green light, confirming that the formation of ROS is also dependent on the intensity of irradiance. The spectral patterns of radicals observed in the cyanobacterial cellular extracts depend completely on the type of species or strain studied and can hardly be generalized; this demonstrates the complexity of the interplay between the formation of ROS, lipid soluble α–tocopherols (vitamin E) and the antioxidant enzyme defence systems (Hsieh et al., 2014). In this case a protective role of OCP and IsiA in the non–photochemical–quenching mechanisms that avoid ROS formation by dissipating excessive light energy as heat could be ruled out since the cells had been treated with methanol and detergent that disrupted the protein–membrane structures.

Photoinhibition

Photoinhibition is the phenomenon observed when the light intensity is beyond the saturation point, where any further increase of light cannot further promote photosynthesis. The latest model of photoinhibition is that ROS generated during photosynthesis may cause photoinhibition by interfering with the elongation step of *psbA* mRNA translation during synthesis *de novo* of D1 protein that is essential for the repair of the photodamaged PSII (Samuelsson et al., 1985; Nishiyama et al., 2004). Photoinhibition can be easily achieved in the dim surroundings of catacombs, underground chambers and caves where the intensity of light is often quite low, usually between 0.2 and 1.7 PAR in the catacombs in Rome (Italy), for example. Therefore, it may be feasible to use low to moderate intensity of light to inhibit the photosynthesis of cyanobacteria, thereby preventing their growth.

The efficiency of light to generate ROS depends not only on its intensity and wavelength, but also on the composition of microbial species in the biofilms, specifically the pigmentation of the microbes. It is known that long wavelengths of light (*e.g.* red light) can easily penetrate into the depth of tissue and reach the cells in the biofilms. Besides, it was observed that red light is especially effective in damaging the bluish green species *Leptolyngbya* sp. and *Scytonema julianum*, which are rich in PC and APC that have λ_{max} in red light region. Green light is effective to inhibit the reddish species *Oculatella subterranea* rich in PE with λ_{max} in green light whereas white light can be used to inhibit the brownish grey *Symphyonemopsis* sp. and the black cyanobacterium *Eucapsis* sp. that have a balanced pigmentation to absorb and be excited by a wide spectrum of light (Hsieh et al., 2014).

Although it was observed that blue light is the most potent generator of ROS in the pigment extracts, it is in fact ineffective to inhibit cyanobacterial growth *in vivo*. The possible explanation may be the protective orange carotenoid protein (OCP). When cyanobacteria are exposed to strong blue–green light, OCP can be triggered by light to change its conformation in a way to attach to the phycobilisomes, thereby transferring the excessive light energy from the phycobilisomes to OCP, from which the excessive light energy is dissipated as heat to the environment (Kirilovsky 2007; Wilson et al., 2007; Karapetyan, 2008; Kirilovsky and Kerfeld, 2013). Since blue light can hardly reach the PSII center, less ROS can actually be generated in living cyanobacteria. For more details about OCP, readers are referred to section 3. (Defence against ROS).

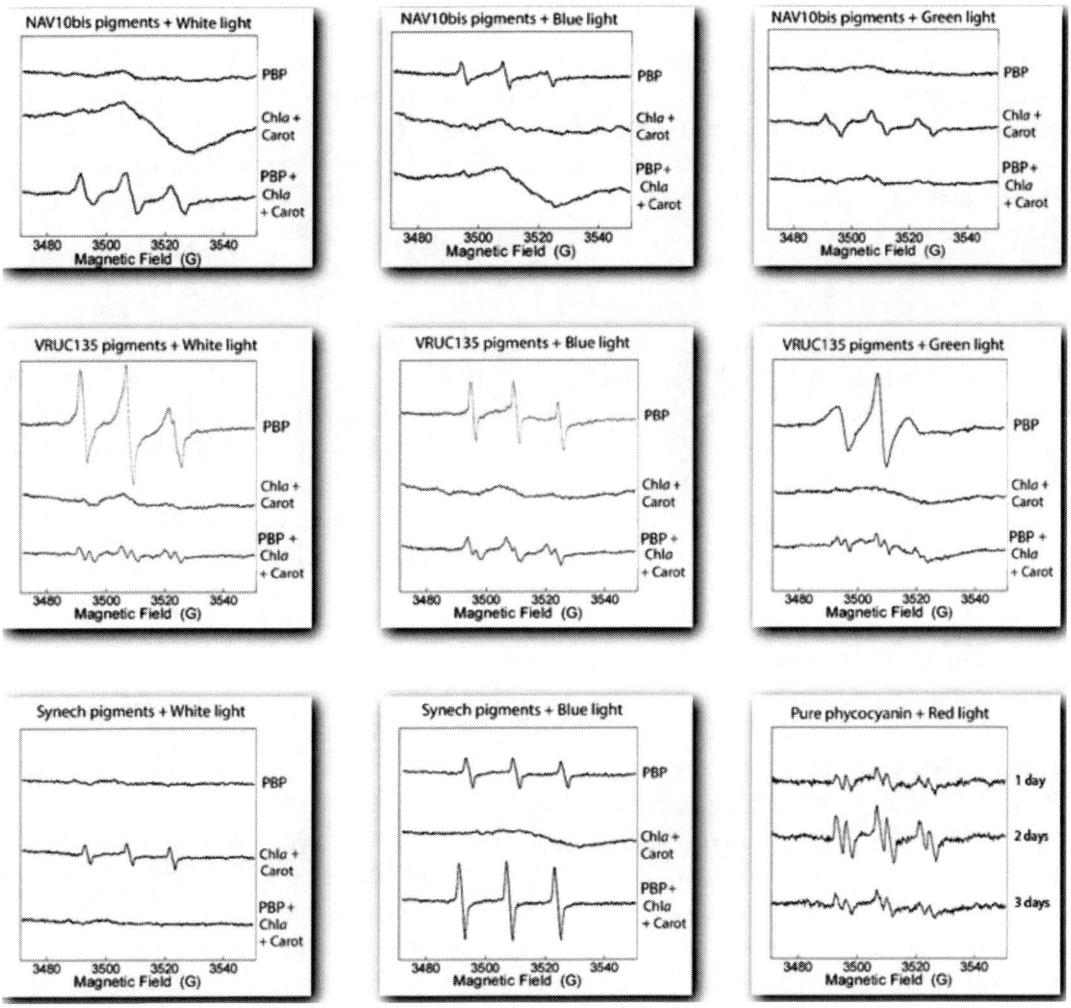

Figure 11. ESR spectroscopy of ROS formation in cyanobacterial cellular extracts irradiated with visible light for 3 days. (Hsieh et al., 2014).

Photobleaching and cellular damage

ROS generated from the photosynthetic pigments can destroy the pigments themselves, a phenomenon known as photobleaching. The fading of pigments was observed with the naked eyes when *Symphyonemopsis* sp. collected from the catacombs was exposed to 50 PAR for 4 days. When 150 PAR was used, the cyanobacterium was bleached into yellowish pale. Replacing these photobleached cells in the dim culture room for 7 days under a 2 PAR white fluorescence tube with 12h/12h dark/light cycle could not restore their pigmentation, implying that *Symphyonemopsis* sp. had suffered severe oxidative cellular damage that could not be repaired within a week (Hsieh et al., 2014).

By using SytoX Green® staining with the aid of confocal laser scanning microscopy, we observed that the majority of the cells in the biofilms exposed to 150 PAR of red, green or white light for 2 weeks had lost their cellular integrity, allowing the penetration of SytoX

Green® into the cells. The results indicate that cyanobacteria within the biofilms could not withstand this lighting condition and had been successfully inactivated by 150 PAR of red, green and white lights. In contrast, 12.5 PAR of red, green and white lights could only cause photoinhibition, but did not result in cellular damage (Hsieh et al., 2014). The results emphasize the importance of light intensity in ROS formation with the aim of preserving cultural heritage from cyanobacterial attack. Blue light could not damage the biofilms, possibly due to its low penetration ability and the protection by OCP.

D–ALA–mediated photodynamic inactivation

The effects of light on ROS generation in cyanobacteria can be enhanced by the addition of δ–Aminolevulinic acid (D–ALA). This is a colorless, small molecule that is a precursor in porphyrin biosynthesis. 8 molecules of D–ALA can be converted into one molecule of tetrapyrrole, such as protoporphyrin IX, Mg–protoporphyrin IX monomethyl ester and protochlorophyllide (Willows, 2003; Walther et al., 2009) (Figure 12). These tetrapyrroles are endogenous photosensitizers that can be excited by appropriate wavelengths of light to produce ROS inside the cells, without the risk of staining or damaging the underlying surface of monuments.

By using ESR spectroscopy, the formation of ROS in the D–ALA treated cyanobacteria was confirmed. 30–40 mM D–ALA produced high levels of ROS in 5 out of 7 isolates tested. The formation of ROS in the D–ALA treated cells is a light–dependent event. After 45 min of red light exposure, for instance, the signal intensity of ROS in D–ALA treated *Leptolyngbya* sp. had already doubled (Hsieh et al., 2013) (Figure 13).

Due to the fact that the microbial biofilms on historical monuments may comprise not only phototrophic cyanobacteria and microalgae, but often also contain heterotrophic actinobacteria and fungi that have no photosynthetic pigments to be excited by light to produce ROS, the application of D–ALA on historical monuments becomes necessary. In fact, D–ALA–mediated photodynamic treatment can inhibit not only cyanobacteria, but also the associated heterotrophic microbes that might thrive after the pioneer cyanobacteria have been eliminated by light.

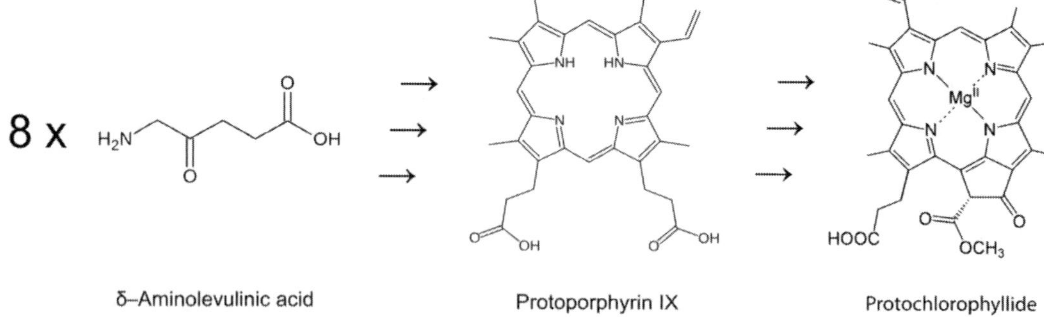

Figure 12. Biosynthesis of tetrapyrroles from D–ALA. A total of 8 molecules of D–ALA produce one molecule of protoporphyrin IX, which is then converted into protochlorophyllide in cyanobacteria.

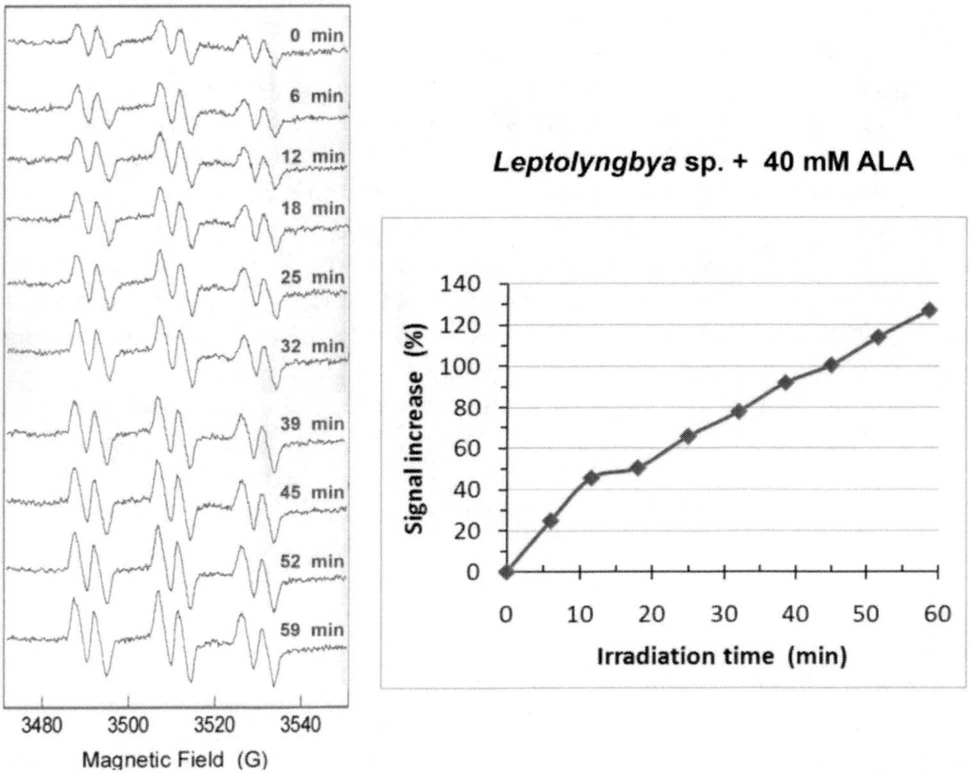

Figure 13. ESR measurements showing the accumulation of ROS in D–ALA–treated *Leptolyngbya* sp. under red light exposure. (Hsieh et al., 2013).

CONCLUSION

ROS in cyanobacteria are continuously produced as normal metabolites in respiration and photosynthesis processes when electrons and light energy are transferred between the excited chlorophyll, PSII, PSI and molecular oxygen. The other pathway of ROS generation is photosensitization of chlorophyll *a* and phycobiliproteins through Type I and Type II photochemical reactions. The production of ROS can be promoted by environmental stress, including high intensity of light, UV radiation, high salinity, drought, either low or medium–high temperatures, iron starvation, CO_2 limitation and other factors that suppress carbon fixation in the Calvin cycle.

ROS have various harmful effects on cyanobacteria. Photoinhibition occurs when the synthesis *de novo* of D1 protein for the repair of the photo–damaged PSII is suppressed by ROS to a rate that cannot compensate the extent of photodamage of the PSII center during photosynthesis. Photobleaching in cyanobacteria follows photoinhibition in the later stages of ROS attack that decompose chlorophyll *a* and phycobiliproteins. DNA breakage is induced by ROS directly through single electron/H–atom reactions and singlet oxygen oxidation, or indirectly through ROS–activated Ca^{+2}–dependent endonucleases. The thylakoid membranes are decomposed through free radical chain reactions when unsaturated fatty acids repeatedly react with ROS and intermediate products. One of the end products, malondialdehyde is a mutagen that forms M_1G-type DNA adducts in cyanobacteria.

Cyanobacteria survive the oxidative stress by using strategies like prevention of ROS formation, scavenging of ROS, and increased expression of photosystem proteins. The protective pigments scytonemin and MAAs in cellular sheath absorb UV radiation and dissipate excess light energy through thermal de–excitation process. The excessive energy from high intensity of light can also be thermally dissipated to the environment by OCP and IsiA through distinct mechanisms. The photomovement ability allows cyanobacteria to escape from damaging sunlight. When high light irradiation cannot be avoided, HLIPs and CAB–like proteins in cyanobacteria can stabilize PSI trimers and chlorophyll–binding proteins that are involved in chlorophyll biosynthesis and tetrapyrrole metabolic pathways.

The global phenomenon of coral bleaching can be attributed to the oxidative stress in symbiotic cyanobacteria. Elevated sea–water temperatures in acidified oceans results in elevated ROS levels in cyanobacteria, which may in turn be expelled by the coral host or simply succumb to the oxidative stress, leading to the whitening of coral reefs. Oxidative stress in cyanobacteria might also explain the seasonal disappearance of algal blooms; the consequences are the release of cyanotoxins and off–odor compounds to the water when cyanobacteria are killed, deteriorating water quality.

However, in the future there could also be positive aspects of the interaction between cyanobacteria and ROS. The most promising water purification method would be the employment of ROS generated from light–activated nano–semiconductors to degrade microcystins, geosmin and 2–methyl–isoborneol. In the field of cultural heritage conservation, ROS could be used as a cleaning tool to rid historical monuments of microbial attack. In terms of bioresource production, ROS may facilitate the generation of electricity in bio–electrochemical systems based on cyanobacterial biocathode, and H_2O_2 produced by cyanobacteria may one day be harvested as a renewable resource from solar energy.

REFERENCES

Al–Rasheed, R. A. (2005). *Water treatment by heterogeneous photocatalysis: a review.* Proceedings of the 4th SWCC acquired experience symposium. Jeddah, Saudi Arabia.

Anderson, J. M., Park, Y.–I. & Chow W. S. (1998). Unifying model for the photoinactivation of Photosystem II *in vivo*: a hypothesis. *Photosynth. Res.*, *56*, 1–13.

Aro, E. M., Virgin, I. & Andersson B. (1993). Photoinhibition of Photosystem II. Inactivation, protein damage and turnover. *Biochim. Biophys. Acta*, *1143*, 113–134.

Asada, K. (1999). The water–water cycle in chloroplasts: scavenging of active oxygens and dissipation of excess photons. *Annu. Rev. Plant Physiol. Plant Mol.*, *50*, 601–639.

Baker A. C., Glynn, P. W. & Riegl, B. (2008). Climate change and coral reef bleaching: an ecological assessment of long–term impacts, recovery trends and future outlook. *Estuar. Coast. Shelf Sci.*, *80*(4), 435–471.

Bamuza–Pemu, E. E. & Chirwa, E. MN (2012). Photocatalytic degradation of geosmin: reaction pathway analysis. *Water SA*, *38*(5), 689–695.

Baulina, O. I., Chivkunova, O. B. & Merzlyak, M. N. (2004). Destruction of pigments and ultrastructural changes in cyanobacteria during photodamage. *Russ. J. Plant Physiol.*, *51*(6), 761–768.

Berera, R., van Stokkum, I. H. M., d'Haene, S., Kennis, J. T. M., van Grondelle, R. & Dekker, J. P. (2009). A mechanism of energy dissipation in cyanobacteria. *Biophys. J.*, *96*, 2261–2267.

Berman–Frank, I., Bidle, K. D., Haramaty, L. & Falkowski, P. G. (2004). The demise of the marine cyanobacterium, *Trichodesmium* spp., via an autocatalyzed cell death pathway. *Limnol. Oceanogr.*, *49*(4), 997–1005.

Bhandari, R. R. & Sharma, P. K. (2010). UV–B radiation and high light induced oxidative damage in *Phormidium corium* may cause bleaching to associated coral reefs. *Indian J. Mar. Sci.*, *39*(3), 423–428.

Bibby, T. S., Mary, I., Nield, J., Partensky, F. & Barber, J. (2003). Low–light–adapted *Prochlorococcus* species posseses specific antennae for each photosystem. *Nature*, *424*, 1051–5054.

Bibby, T. S., Nield, J. & Barber, J. (2001). Iron deficiency induces formation of an antenna ring around trimeric photosystem I in cyanobacteria. *Nature*, *412*, 743-745.

Billi, D. (2009). Subcellular integrities in *Chroococcidiopsis* sp. CCMEE 029 survivors after prolonged desiccation revealed by molecular probes and genome stability assays. *Extremophiles*, *13*, 49–57.

Billi, D. & Potts, M. (2002). Life and death of dried prokaryotes. *Res. Microbiol.*, *153*, 7–12.

Boekema, E. J., Higney, A., Yakushevska, A. E., Piotrowski, M., Keegstra, W., Berry, S., Marcel, K.-P., Pistorius, E. K. & Kruip, J. (2001). A giant chlorophyll-protein complex induced by iron deficiency in cyanobacteria. *Nature*, *412*, 745–748.

Borg, D. C., Schaich, K. M., Jr. Elmore, J. J. & Bell, J. A. (1978). Cytotoxic reactions of free radical species of oxygen. *Photochem. Photobiol.*, *28*, 887–907.

Bouchard, J. N. & Purdie, D. A. (2011). Effect of elevated temperature, darkness and hydrogen peroxide treatment on oxidative stress and cell death in the bloom–forming toxic cyanobacterium *Microcystis aeruginosa*. *J. Phycol.*, *47*, 1316–1325.

Boulay, C., Abasova, L., Six, C., Vass, I. & Kirilovsky, D. (2008). Occurrence and function of the orange carotenoid protein in photoprotective mechanisms in various cyanobacteria. *Biochim. Biophys. Acta*, *1777*, 1344–1354.

Britton, G., Liaaen–Jensen, S. & Pfander, H. (Eds). Carotenoids Handbook. Birkhäuser, Basel, Switzerland (2008). 660 pp.

Brocks, J. J., Logan, G. A., Buick, R. & Summons, R. E. (1999). Archean molecular fossils and the early rise of eukaryotes. *Science*, *285*, 1033–1036.

Büdel, B. & Wessels, D. C. J. (1991). Rock inhibiting blue–green algae/cyanobacteria from hot arid regions. *Algol. Stud. 64*, 385–398.

Cadet, J., Ravanat, J. L., Martinez, G. R., Medeiros, M. H. & Di Mascio, P. (2006). Singlet oxygen oxidation of isolated and cellular DNA: product formation and mechanistic insights. *Photochem. Photobiol.*, *82*, 1219–1225.

Cadet, J. & Téoule, R. (1978). Comparative study of oxidation of nucleic acid components by hydroxyl radicals, singlet oxygen and superoxide anion radical. *Photochem. Photobiol.*, *28*, 661–667.

Cai, P.–J., Xiao, X., He, Y.–R., Li, W.–W., Zang, G.–L., Sheng, G.–P., Lam, M. H.–W., Yu, L & Yu, H.–Q. (2013). Reactive oxygen species (ROS) generated by cyanobacteria act as an electron acceptor in the biocathode of a bio–electrochemical system. *Biosens. Bioelectro.*, *39*, 306–310.

Callahan, F. E., Becker, D. W. & Cheniae, G. M. (1986). Studies on the photoactivation of the water–oxidizing enzyme: II. Characterization of weak light photoinhibition of PSII and its light–induced recovery. *Plant Physiol.*, *82*(1), 261–269.

Cameron, J. C. & Pakrasi, H. B. (2010). Essential role of glutathione in acclimation to environmental and redox perturbations in the cyanobacterium *Synechocystis* sp. PCC 6803. *Plant Physiol.*, *154*, 1672–1685.

Campos–Martin, J. M, Blanco–Brieva, G. & Fierro, J. L. G. (2006). Hydrogen peroxide synthesis: an outlook beyond the antraquinone process. *Angewandte Chem. Internl. Ed.*, *45*(42), 6962–6984.

Canini, A., Leonardi, D. & Grilli Caiola, M. (2001). Superoxide dismutase activity in the cyanobacterium *Microcystis aeruginosa* after surface bloom formation. *New Phytol.*, *21*, 107–116.

Carrette, L., Friedrich, K. A. & Stimming, U. (2000). Fuel cells: principles, types, fuels, and applications. *ChemPhysChem*, *1*(4), 162–193.

Castenholz, R. W. & Garcia–Pichel, F. (2012). Cyanobacterial responses to UV radiation. *In*: B. A. Whitton (Ed.) Ecology of Cyanobacteria II: Their Diversity in Space and Time, 317–343. Springer, Dordrecht (The Netherlands), 760 pp.

Chaneva, G., Petrova, D. & Uzunova, A. (2009). Changes in antioxidative enzymes of a cyanobacterium grown under heavy metal stress. *Biotechnol. Biotechnol. Equip.*, *23*, 554-556.

Chauhan D., Folea, I. M., Jolley, C. C., Kouřil, R., Lubner, C. E., Lin, S., Kolber, D., Wolfe–Simon, F., Golbeck, J. H., Boekema, E. J. & Fromme, P. (2011). A novel photosynthetic strategy for adaptation to low-iron aquatic environments. *Biochemistry*, *50*, 686–692.

Cockell, C. S., Schuerger, A. C., Billi, D., Friedmann, E. I. & Panitz, C. (2005). Effects of a simulated Martian UV flux on the cyanobacterium, *Chroococcidiopsis sp. 029. Astrobiology, 5*(2), 127–140.

Conde, F. R., Churio, M. S. & Previtali, C. M. (2004). The deactivation pathways of the excited–states of the mycosporine-like amino acids shinorine and porphyra–334 in aqueous solution. *Photochem. Photobiol. Sci.*, *3*, 960–967.

Czapski, G. (1984). Reaction of OH•. *Methods Enzymol.*, *105*, 209–215.

Davies, M. J. (2004). Reactive species formed on proteins exposed to singlet oxygen. *Photochem. Photobiol. Sci.*, *3*, 17–25.

Dietz, K.–J. (2010). Peroxiredoxins in plants and cyanobacteria. *Antiox. Redox Signal.*, *15*, 1129–1159.

Di Mascio, P., Murphy, M. E. & Sies, H. (1991). Antioxidant defence systems: the role of carotenoids, tocopherols and thiols. *Am. J. Clin. Natr.*, *53*, 194S–200S.

Ding, Y., Gan, N., Li, J., Sedmak, B. & Song, L. (2012). Hydrogen peroxide induces apoptotic–like cell death in *Microcystis aeruginosa* (Chroococcales, Cyanobacteria) in a dose–dependent manner. *Phycologia*, *51*(5), 567–575.

Dionisio–Sese, M. L. (2010). Aquatic microalgae as potential sources of UV–screening compounds. *Philip. J. Sci.*, *139*, 5–16.

Dolmans, D. E. J. G. J., Fukumura, D. & Jain, R. K. (2003). Photodynamic therapy for cancer. *Nature Rev. Cancer*, *3*, 380–387.

Donkor, V., Amewowor, D. & Häder, D.–P. (1993). Effects of tropical solar radiation on the motility of filamentous cyanobacteria. *FEMS Microbiol. Ecol.*, *12*, 143–148.

Eckert, H. J., Geiken, B., Bernarding, J., Napiwotzki, A., Eichler, H. J. & Renger, G. (1991). Two sites of photoinhibition of the electron–transfer in oxygen evolving and Tris–treated PSII membrane fragments from spinach. *Photosynth. Res.*, *27*(2), 97–108.

Elstner, E. F. & Pils, I. (1979). Ethane formation and chlorophyll bleaching in DCMU–treated *Euglena gracilis* cells and isolated spinach chloroplast lamellae. *Z. Naturforsch 34c*, 1040–1043.

Feller, U., Crafts–Brandner, S. J. & Salvucci, M. E. (1998). Moderate high temperatures inhibit ribulose–1,5–bisphosphate carboxylase/oxygenase (Rubisco) activase–mediated activation of Rubisco. *Plant Physiol.*, *166*, 539–546.

Flewelling, L. J., Naar, J. P., Abbott, J. P., Baden, D. G., Barros, N. B., Bossart, G. D., Bottein, M.–Y. D., Hammond, D. G., Haubold, E. M., Heil, C. A., Henry, M. S., Jacocks, H. M., Leighfield, T. A., Pierce, R. H., Pitchford, T. D., Rommel, S. A., Scott, P. S., Steidinger, K. A., Truby, E. W., Van Dolah, F. M. & Landsberg, J. H. (2005). Red tides and marine mammal mortalities: unexpected brevetoxin vectors may account for deaths long after or remote from an algal bloom. *Nature*, *435*(7043), 755–756.

Fotious, T., Triantis, T. M., Kaloudis, T., Pastrana–Martínez, L. M., Likodimos, V., Falaras, P., Silva, A. M. T. & Hiskia, A. (2013). Photocatalytic degradation of microcystin–LR and off–odor compounds in water under UV–A and solar light with a nanostructured photocatalyst based on reduced grapheme oxide–TiO_2 composite. Identification of intermediate products. *Indust. Eng. Chem. Res.*, *52*(39), 13991–14000.

Frangeul, L., Quillardet, P., Castets, A. M., Humbert, J. F., Matthijs, H. C. P., Cortez, D., Tolonen, A., Zhang, C. C., Gribaldo, S., Kehr, J. C., Zilliges, Y., Ziemert, N., Becker, S., Talla, E., Latifi, A., Billault, A., Lepelletier, A., Dittmann, E., Bouchier, R. & De Marsac, N.T. (2008). Highly plastic genome of *Microcystis aeruginosa* PCC 7806, a ubiquitous toxic freshwater cyanobacterium. *BMC Genomics*, *9*, 274, doi:10.1186/1471-2164-9-274.

Fridovich, I. (1998). Oxygen toxicity: a radical explanation. *J. Exp. Biol.*, *201*, 1203–1209.

Friedmann, E. I. & Ocampo–Friedmann, R. (1985). Blue–green algae in arid cryptoendolithic habitats. *Arch. Hydrobiol.*, 38/39, 349–350.

Funk, C. & Vermaas, W. (1999). A cyanobacterial gene family coding for single–helix proteins resembling part of the light–harvesting proteins from higher plants. *Biochemistry*, *38*, 9397–9404.

Garcia–Pichel, F. (1994). A model for internal self–shading in planktonic organisms and its implications for the usefulness of ultraviolet sunscreens. *Limnol. Oceanogr.*, *39*(7), 1704–1717.

Garcia–Pichel, F., Mechling, M. & Castenholz, R. W. (1994). Diel migrations of microorganisms within a benthic, hypersaline mat community. *Appl. Environ. Microbiol.*, *60*, 1500–1511.

Garcia–Pichel, F. & Castenholz, R. W. (1991). Characterization and biological implications of scytonemin, a cyanobacterial sheath pigment. *J. Phycol.*, *27*, 395–409.

Gaya, U. I. & Abdullah, A. H. (2008). Heterogeneous photocatalytic degradation of organic contaminants over titanium dioxide: a review of fundamentals, progress and problems. *J. Photochem. Photobiol. C: Photochem. Rev.*, *9*(1), 1–12.

Giese, B. (2000). Long–distance charge transfer in DNA: the hopping mechanism. *Acc. Chem. Res.*, *33*, 631–636.

Gladis, F., Eggert, A., Karasten, U. & Schumann, R. (2010). Prevention of biofilm growth on man–made surfaces: evaluation of antialgal activity of two biocides and photocatalytic nanoparticles. *Biofouling*, *26*(1), 89–101.

Grilli Caiola, M. & Billi, D. (2007). *Chroococcidiopsis* from desert to Mars. *In*: J. Seckbach (Ed). Algae and cyanobacteria in extreme environments. Cellular origin, life in extreme habitats and astrobiology. Springer, Berlin, 553–568.

Hakala, M., Tuominen, I., Keränen, M., Tyystjärvi, T. & Tyystjärvi, E. (2005). Evidence for the role of the oxygen-evolving manganese complex in photoinhibition of Photosystem II. *Biochim. Biophys. Acta*, *1706*(1–2), 68–80.

Halliwell, B. & Gutteridge, J. M. C. (1990). Role of free radicls and catalytic metal ions in human disease: an overview. *Methods Enzymol.*, *186*, 1–85.

Halliwell, B. & Gutteridge, J. M. C. (Eds). *Free Radicals in Biology and Medicine*. Oxford, New York (1999).

Havaux, M., Guedeney, G., Hagemann, M., Yeremenko, N., Matthijs, H. C. P. & Jeanjean, R. (2005). The chlorophyll–binding protein IsiA is inducible by high light and protects the cyanobacterium *Synechocystis* PCC6803 from photooxidative stress. *FEMB Lett.*, *579*, 2289–2293.

He, J.–A., Hu, Y.–Z. & Jiang, L.–J. (1997). Photodynamic action of phycobiliproteins: in situ generation of reactive oxygen species. *Biochim. Biophysic. Acta*, *1320*(2), 165–174.

He, Q., Dolganov, N., Björkman, O. & Grossman, A. R. (2001). The high light–inducible polypeptides in *Synechocystis* PCC 6803. *J. Biol. Chem.*, *276*(1), 306–314.

He, Y.–Y. & Häder, D.–P. (2002a). Involvement of reactive oxygen species in the UV–B damage to the cyanobacterium *Anabaena* sp. *Photochem. Photobiol.*, *66*, 73–80.

He, Y.–Y. & Häder, D.–P. (2002b). UV–B–induced formation of reactive oxygen species and oxidative damage of the cyanobacterium *Anabaena* sp.: protective effects of ascorbic acid and N–acetyl–L–cysteine. *J. Photochem. Photobiol. B*, *66*, 115–124.

Hernández–Prieto, M. A., Tibiletti, T., Abasova, L., Kirilovsky, D., Vass, I. & Funk, C. (2011). The small CAB–like proteins of cyanobacterium *Synechocystis* sp. PCC 6803: their involvement in chlorophyll biogenesis for Photosystem II. *Biochim. Biophys. Acta*, *1807*, 1143–1151.

Hoegh–Guldberg, O. (1999). Climate change, coral bleaching and the future of the world's coral reefs. *Mar. Freshw. Res.*, *50*(8), 839–866.

Hoegh–Guldberg, O., Mumby, P. J., Hooten, A. J., Steneck, R. S., Greenfield, P., Gomez, E., Harvell, C. D., Sale, P. F., Edwards, A. J., Caldeira, K., Knowlton, N., Eakin, C. M., Iglesias–Prieto, R., Muthiga, N., Bradbury, R. H., Dubi, A. & Hatziolos, M. E. (2007). Coral reefs under rapid climate change and ocean acidification. *Science*, *318*, 1737–1742.

Holt, N. E., Zigmantas, D., Valkunas, L., Li, X.–P., Niyogi, K. K. & Fleming, G. R. (2005). Carotenoid cation formation and the regulation of photosynthetic light harvesting. *Science*, *307*, 433–436.

Holt, T. K. & Krogmann, D. W. (1981). A carotenoid protein from cyanobacteria. *Biochim. Biophys. Acta*, *637*, 408–414.

Hosoya–Matsuda, N., Motohashi, K., Yoshimura, H., Nozaki, A., Inoue, K., Ohmori, M. & Hisabori, T. (2005). Anti–oxidative stress system in cyanobacteria. Significance of type II peroxiredoxin and the role of 1–Cys peroxiredoxin in *Synechocystis* sp. strain PCC 6803. *J. Biol. Chem.*, *280*, 840–846.

Hsieh, P., Pedersen, J. Z. & Albertano, P. (2013). Generation of reactive oxygen species upon red light exposure of cyanobacteria from Roman hypogea. *Internl. Biodeter. Biodegrad.*, *84*, 258–265.

Hsieh, P., Pedersen, J. Z. & Bruno, L. (2014). Photoinhibition of cyanobacteria and its application in cultural heritage conservation. *Photochem. Photobiol.*, *90*, 533–543.

Ivanov, A. G., Park, Y. I., Miskiewicz, E., Raven, J. A., Huner, N. P. & Oquist, G. (2000). Iron stress restricts photosynthetic intersystem electron transport in *Synechocystis* sp. PCC 7942. *FEBS Lett.*, *485*, 173–177.

Johnson, M., Havaux, M., Triantaphylidès, C., Ksas, B., Pascal, A. A., Robert, B., Davidson, P. A., Ruban, A. V. & Horton, P. (2007). Elevated zeaxanthin bound to oligomeric LHCII enhances the resistance of *Arabidopsis* to oxidative stress by a lipid–protective, antioxidant mechanism. *J. Biol. Chem.*, *282*, 22605–22618.

Jones, C. M., Lawrence, A., Wardman, P. & Burkitt, M. J. (2003). Kinetics of superoxide scavenging by glutathione: an evaluation of its role in the removal of mitochondrial superoxide. *Biochem. Soc. Trans.*, *31*, 1337–1339.

Jung, J. & Kim, H.–S. (1990). The chromophores as endogenous sensitizers involved in the photogeneration of singlet oxygen in spinach thylakoids. *Photochem. Photobiol.*, *52*(5), 1003–1009.

Kaneko, T., Nakajima, N., Okamoto, S., Suzuki, I., Tanabe, Y., Tamaoki, M., Nakamura, Y., Kasai, F., Watanabe, A., Kawashima, K., Kishida, Y., Ono, A., Shimizu, Y., Takahashi, C., Minami, C., Fujishiro, T., Kohara, M., Katoh, M., Nakanaki, N., Nakayama, S., Yamada, M., Tabata, S. & Watanabe, M. M. (2007). Complete genomic structure of the bloom–forming toxic cyanobacterium *Microcystis aeruginosa* NIES–843. *DNA Res.*, *14*, 247–256.

Karapetyan, N. V. (2008). Protective dissipation of excess absorbed energy by photosynthetic apparatus of cyanobacteria: role of antenna terminal emitters. *Photosynth. Res.*, *97*, 195–204.

Keren, N., Berg, A., van Kan, P. J. M., Levanon, H. & Ohad, I. (1997). Mechanism of Photosystem II photoinactivation and D1 protein degradation at low light: the role of back electron flow. *Proc. Natl. Acad. Sci. USA*, *94*(4), 1579–1584.

Kerfeld, C. A., Sawaya, M. R., Brahmandam, V., Cascio, D., Ho, K. K., Trevithick–Sutton, C. C., Krogmann, D. W. & Yeates, T. O. (2003). The crystal structure of a cyanobacterial water–soluble carotenoid biding protein. *Structure*, *11*, 55–66.

Kim, S.–C. & Lee, D.–K. (2005). Preparation of TiO_2–coated hollow glass beads and their application to the control of algal growth in eutrophic water. *Microchem. J. 80*, 277–232.

Kim, J. H. & Suh, K. H. (2005). Light–dependent expression of superoxide dismutase from cyanobacterium *Synechocystis* sp. Strain PCC 6803. *Arch. Microbiol.*, *183*, 218–223.

Kirilovsky, D. (2007). Photoprotection in cyanobacteria: the orange carotenoid protein (OCP)–related non–photochemical–quenching mechanism. *Photosynth. Res.*, *93*, 7–16.

Kirilovsky, D. & Kerfeld, C. A. (2013). The orange carotenoid protein: a blue–green light photoactive protein. *Photochem. Photobiol. Sci.*, *12*, 1135–1143.

Kojima, K., Suzuki–Maenaka, T., Kikuchi, T. & Nakamoto, H. (2006). Roles of the cyanobacterial isiABC operon in protection from oxidative and heat stresses. *Physiol. Plant.*, *128*, 507–519.

Kruschel, C., Richard, W. & Castenholz, W. (1998). The effect of solar UV and visible irradiance on the vertical movements of cyanobacteria in microbial mats of hypersaline waters. *FEMS Microbiol. Ecol.*, *27*, 53–72.

Kumar, A., Tyagi, M. B., Singh, N., Tyagi, R., Jha, P. N., Sinha, R. P. & Häder, D.–P. (2003). Role of white light in reversing UV–B–mediated effects in the N_2–fixation cyanobacterium *Anabaena* BT2. *J. Photochem. Photobiol. B, 71*, 35–42.

Latifi, A., Jeanjean, R., Lemeille, S., Havaux, M. & Zhang, C.–C. (2005). Iron starvation leads to oxidative stress in *Anabaena* sp. PCC 7120. *J. Bacteriol.*, *187*(18), 6596–6598.

Leverenz, R. L., Jallet, D., Li, M.–D., Mathies, R. A., Kirilovsky, D. & Kerfeld, C. A. (2014). Structural and functional modularity of the orange carotenoid protein: distinct roles for the N– and C–terminal domains in cyanobacterial photoprotection. *Plant Cell*, *26*, 426–437.

Loya, Y., Sakai, K., Yamazato, K., Nakano, Y., Sambali, H. & van Woesik, R. (2001). Coral bleaching: the winners and the losers. *Eol. Lett.*, *4*, 122–131.

Marnett, L. J. (1999). Lipid peroxidation–DNA damage by malondialdehyde. *Mutat. Res.*, *424*(1–2), 83–95.

Martin, J. & Fitzwater, S. (1988). Iron deficiency limits phytoplankton growth in the north–east Pacific subarctic. *Nature*, *331*, 341–343.

Mattimore, V. & Battista, J. R. (1996). Radioresistance of *Deinococcus radiodurans*: functions necessary to survive ionizing radiation are also necessary to survive prolonged desiccation. *J. Bacteriol.*, *178*, 633–637.

McGuire, M. J. (1995). Off-flavor as the consumer's measure of drinking water safety. *Water Sci. Technol.*, *31*(11). 1.

Mehler, A. H. (1951a). Studies on reactions of illuminated chloroplasts. I. Mechanism of the reduction of oxygen and other Hill reagents. *Arch. Biochem. Biophysic.*, *33*(1), 65–77.

Mehler, A. H. (1951b). Studies on reactions of illuminated chloroplasts. II. Stimulation and inhibition of the reaction with molecular oxygen. *Arch. Biochem. Biophysic. 34*(2), 339–351.

Murata, N., Takahashi, S., Nishiyama, Y. & Allakhverdiev, S. I. (2007). Photoinhibition of Photosystem II under environmental stress. *Biochim Biophys. Acta*, *1767*, 414–421.

Ning, S. B., Gao, H. L., Wang, L. & Song, Y. C. (2002). Salt stress induces programmed cell death in prokaryotic organism *Anabaena*. *J. Appl. Microbiol.*, *93*, 15–28.

Nishiyama, Y., Allakhverdiev, S. I. & Murata, N. (2006). A new paradigm for the action of reactive oxygen species in the photoinhibition of photosystem II. *Biochim. Biophys. Acta*, *1757*, 742–749.

Nishiyama, Y., Allakhverdiev, S. I., Yamamoto, H., Hayashi, H. & Murata, N. (2004). Singlet oxygen inhibits the repair of Photosystem II by suppressing translation elongation of the D1 protein in *Synechocystis* sp. PCC 6803. *Biochemistry*, *43*, 11321–11330.

Nixon, P. J., Michoux, F., Yu, J., Boehm, M., & Komenda, J. (2010). Recent advances in understanding the assembly and repair of photosystem II. *Annals Botany*, *106*, 1–16.

Ohko, Y., Nagao, Y., Okano, K., Sugiura, N., Fukuda, A., Yang, Y., Negishi, N., Takeuchi, M. & Hanada, S. (2009). Prevention of *Phormidium tenue* biofilm formation by TiO_2 photocatalysis. *Microb. Environ.*, *24*(3), 241–245.

Ohnishi, N., Allakhverdiev, S. I., Takahashi, S., Higashi, S., Watanabe, M., Nishiyama, Y. & Murata, N. (2005). Two–step mechanism of photodamage to Photosystem II: Step 1 occurs at the oxygen–evolving complex and step 2 occurs at the photochemical reaction center. *Biochemistry, 44*(23), 8494–8499.

Paul, B. T., Patel, A., Selvam, G. S., Mishra, S., Ghosh, P. K. & Murugesan, R. (2006). Photodynamic action of C–phycocyanins obtained from marine and fresh water cyanobacterial cultures: a comparative study using EPR spin trapping technique. *Free Radic. Res., 40*(8), 821–825.

Potts, M. (1994). Mechanisms of desiccation tolerance in cyanobacteria. *Eur. J. Phycol., 34*, 319–328.

Price, A. H., Atherton, N. M. & Hendry, G. A. F. (1989). Plants under drought stress generate activated oxygen. *Free Radic. Res. Commun., 8*, 61–66.

Price, A. H. & Hendry, G. A. F. (1991). Iron–catalyzed oxygen radical formation and its possible contribution to drought damage in nine native grasses and three cereals. *Plant Cell Environ., 14*, 477–484.

Priya, B., Premanandh, J., Dhanalakshmi, R. T., Seethalakshmi, T., Lakshmanan Uma, L., Prabaharan, D. & Subramanian, G. (2007). Comparative analysis of cyanobacterial superoxide dismutases to discriminate canonical forms. *BMC Genom., 8*, 435.

Priya, B., Sivaprasanth, R. K., Jensi, V. D., Lakshmanan Uma, L., Subramanian, G. & Prabaharan, D. (2010). Characterization of manganese superoxide dismutase from a marine cyanobacterium *Leptolyngbya valderiana* BDU20041. *Sal. Syst., 6*, 6.

Proteau, P. J., Gerwick, W. H., Garcia–Pichel, F. & Castenholz, R. W. (1993). The structure of scytonemin, an ultraviolet sunscreen pigment from the sheaths of cyanobacteria. *Experientia, 49*, 825–829.

Radmer, R. & Ollinger, O. (1980). Light–driven uptake of oxygen, carbon dioxide, and bicarbonate by the green alga *Scenedesmus, Plant Physiol., 65*, 723–729.

Raoul, S. & Cadet, J. (1996). Photosensitized reaction of 8–oxo–7,8–dihydro–2'–deoxyguanosine: identification of 1–(2–deoxy–β–D–*erythro*–pentofuranosyl)cyanuric acid as the major singlet oxygen oxidation product. *J. Am. Chem. Soc., 118*, 1892–1898.

Ravanat, J.–L., Douki, T. & Cadet, J. (2001). Direct and indirect effects of UV radiation on DNA and its components. *J. Photochem. Photobiol. B: Biology, 63*, 88–102.

Redmond, R. W. & Kochevar, I. E. (2006). Spatially resolved cellular responses to singlet oxygen. *Photochem. Photobiol., 82*, 1178–1186.

Rinalducci, S., Hideg, É., Vass, I. & Zolla, L. (2006). Effect of moderate UV–B irradiation on *Synechocystis* PCC 6803 biliproteins. *Biochem. Biophys. Res. Commun., 341*, 1105–1112.

Rinalducci, S., Pedersen, J. Z., & Zolla, L. (2004). Formation of radicals from singlet oxygen produced during photoinhibition of isolated light–harvesting proteins of Photosystem II. *Biochim. Biophys. Acta, 1608*, 63–73.

Rinalducci, S., Pedersen, J. Z., & Zolla, L. (2008). Generation of reactive oxygen species upon strong visible light irradiation of isolated phycobilisomes from *Synechocystis* PCC 6803. *Biochim. Biophys. Acta, 1777*, 417–424.

Roncel, M., Navarro, J. A. & De La Rosa, M. A. (1989). Coupling of solar energy to hydrogen peroxide production in the cyanobacterium *Anacystis nidulans. App. Environ. Microbiol., 55*(2), 483–487.

Ross, C., Santiago-Vázquez, L. & Paul, V. (2006). Toxin release in response to oxidative stress and programmed cell death in the cyanobacterium *Microcystis aeruginosa. Aquat. Toxicol. 78*, 66–73.

Ryan–Keogh, T.J. & Macey, A. I. (2012). The cyanobacterial chlorophyll–binding–protein IsiA acts to increase the *in vivo* effective absorption gross-section of PSI under iron limitation. *J. Phycol.*, *48*, 145–154.

Ryter, S. W., Kim, H. P., Hoetzel, A., Park, J. W., Nakahira, K., Wang, X. & Choi, A. M. K. (2007). Mechanisms of cell death in oxidative stress. *Antioxid. Redox Signal.*, *9*(1), 50–89.

Samuelsson, G, Lönneborg, A., Rosenqvist, E., Gustafsson, P, & Öquist, G. (1985). Photoinhibition and reactivation of photosynthesis in the cyanobacterium *Anacystis nidulans*. *Plant Physiol.*, *79*, 992–995.

Sandström, S. Park, Y. I., Oquist, G. & Gustafsson, P. (2001). CP43', the isiA gene product, functions as an excitation energy dissipator in the cyanobacterium *Synechococcus* sp. PCC 7942. *Photochem. Photobiol.*, *74*, 431–437.

Santabarbara, S., Cazzalini, I., Rivadossi, A., Garlaschi, F. M., Zucchelli, G. & Jennings, R. C. (2002). Photoinhibition *in vivo* and *in vitro* involves weakly coupled chlorophyll-protein complexes. *Photochem. Photobiol.*, *75*(6), 613–618.

Schmetterer, G. (1982). Formation of hydrocarbons by photo–bleaching cyanobacterium *Anacystis nidulans*. *Z. Naturforsch.*, *37c*, 205–209.

Schmetterer, G., Peschek, G. A. & Sleytr, U. B. (1983). Thylakoid degradation during photooxidative bleaching of the cyanobacterium *Anacystis nidulans*. *Protoplasma*, *115*, 202–207.

Schuster, G. B. (2000). Long–range charge transfer in DNA: transient structural distortions control the distance dependence. *Acc. Chem. Res.*, *33*, 253–260.

Senaratna, T. & McKersie, B. D. (1986). Loss of desiccation tolerance during seed germination: a free radical mechanism of injury. *In*: Membranes, Metabolism and Dry Organisms. A.C. Leopold (Ed). Cornell University Press, N.Y. 85–101.

Sheu, C. & Foote, C. S. (1993). Endoperoxide formation in a guanosine derivative. *J. Am. Chem. Soc.*, *115*, 10446–10447.

Shirkey, B., Kovarcik, D. P., Wright, D. J., Wilmoth, G., Prickett, T. F., Helm, R. F., Gregory, E. M. & Potts, M. (2000). Active Fe-containing superoxide dismutase and abundant sodF mRNA in *Nostoc commune* (cyanobacteria) after years of desiccation. *J. Bact.*, *182*, 189–197.

Shirkey, B., McMaster, N. J., Smith, S. C., Wright, D. J., Rodriguez, H., Jaruga, P., Birincioglu, M., Helm, R.F. & Potts, M. (2003). Genomic DNA of *Nostoc commune* (cyanobacteria). becomes covalently modified during long–term (decades) desiccation but is protected from oxidative damage and degradation. *Nucleic Acids Res.*, *31*, 2995–3005.

Sinha, R. P., Klisch, M., Gröniger, A. & Häder, D. P. (2001). Responses of aquatic algae and cyanobacteria to solar UV–B. *Plant Ecol.*, *154*, 219–236.

Slater, A. F. G., Nobel, C. S. I. & Orrenius, S. (1995). The role of intracellular oxidants in apoptosis. *Biochim. Biophys. Acta*, *1271*, 59–62.

Sobotka, R., McLean, S., Zuberova, M., Hunter, C. N. & Tichy, M. (2008). The C–terminal extension of ferrochelatase is critical for enzyme activity and for functioning of the tetrapyrroles pathway in *Synechocystis* PCC 6803. *J. Bacteriol.* *190*(6), 2086–2095.

Solomon, A., Beer, S., Waisel, Y., Jones, G. P. & Paleg, L. G. (1994). Effects of NaCl on the carboxylating activity of Rubisco from *Tamarix jordanis* in the presence and absence of proline–related compatible solutes. *Physiol. Plant*, *90*, 198–204.

Soule, T., Stout, V., Swingley, W. D., Meeks, J. C. & Garcia–Pichel, F. (2007). Molecular genetics and genomic analysis of scytonemin biosynthesis in *Nostoc punctiforme* ATCC 29133. *J. Bacteriol. 189*, 4465–4472.

Stevens, S. E., Patterson, C. O. P. & Myers, J. (1973). The production of hydrogen peroxide by blue–green algae: a survey. *J. Phycol.*, *9*, 427–430.

Suh, H.–J., Lee, H.–W. & Jung, J. (2003). Mycosporine glycine protects biological systems against photodynamic damage by quenching singlet oxygen with a high efficiency. *Photochem. Photobiol.*, *78*(2), 109–113.

Takahashi, S. & Murata, N. (2008). How do environmental stresses accelerate photoinhibition? *Trends Plant Sci.*, *13*(4), 178–182.

Takaichi, S. & Mochimaru, M. (2007). Carotenoids and carotenogenesis in cyanobacteria: unique ketocarotenoids and carotenoid glycoside. *Cell. Mol. Life Sci.*, *64*, 2607–2619.

Tamulonis, C., Postma, M. & Kaandorp, J. (2011). Modeling filamentous cyanobacteria reveals the advantages of long and fast trichomes for optimizing light exposure. *PloS ONE*, *6*(7), e22084, 1–12.

Tel–Or, E., Huflejt, M. & Packer, L. (1985). The role of glutathione and ascorbate in hydroperoxide removal in cyanobacteria. *Biochem. Biophys. Res. Commun.*, *132*, 533–539.

Tel–Or, E., Huflejt, M. E. & Packer, L. (1986). Hydroperoxide metabolism in cyanobacteria. *Arch. Biochem. Biophys.*, *246*, 396–402.

Thomas, D. J., Avenson, T. J., Thomas, J. B. & Herbert, S. K. (1998). A cyanobacterium lacking iron superoxide dismutase is sensitized to oxidative stress induced with methyl viologen but is not sensitized to oxidative stress induced with norflurazon. *Plant Physiol.*, *116*, 1593–1602.

Triantaphylidès, C. & Havaux, M. (2009). Singlet oxygen in plants: production, detoxification and signaling. *Trends Plant Sci.*, *14*, 219–228.

Triantaphylidès, C., Krischke, M., Hoeberichts, F. A., Ksas, B., Gresser, G., Havaux, M., Van Breusegem, F. & Mueller, M.J. (2008). Singlet oxygen is the major reactive oxygen species involved in photo–oxidative damage to plants. *Plant Physiol.*, *148*, 960–968.

Tyystjärvi, E. (2008). Photoinhibition of Photosystem II and photodamage of the oxygen–evolving manganese cluster. *Coordin. Chem. Rev.*, *252*(3–4), 361–376.

Vass, I., Kirilovsky, D. & Etienne, A. L. (1999). UV–B radiation–induced donor–and–acceptor-side modifications of photosystem II in the cyanobacterium *Synechocystis* sp. PCC 6803. *Biochemistry*, *38*, 12786–12794.

Vass, I., Styring, S., Hundal, T., Koivuniemi, M., Aro, E.–N. & Andersson, B. (1992). Reversible and irreversible intermediates during photoinhibition of photosystem II: Stable reduced Q_A species promote chlorophyll triplet formation. *Proc. Natl. Acad. Sci. USA*, *89*(4), 1408–1412.

Vavilin, D., Yao, D. & Vermaas, W. (2007). Small Cab–like proteins retard degradation of Photosystem II-associated chlorophyll in *Synechocystis* sp. PCC 6803: kinetic analysis of pigment labeling with ^{15}N and ^{13}C. *J. Biol. Chem.*, *282*(52), 37660–37668.

Vinnemeier, J., Kunert, A. & Hagemann, M. (1998). Transcriptional analysis of the isiAB operon in salt–stressed cells of the cyanobacterium *Synechocystis* sp. PCC 6803. *FEMS Microbiol. Lett.*, *169*, 323–330.

Walther, J., Bröcker, M. J., Wätzlich, D., Nimtz, M., Rohde, M., Jahn, D. & Moser, J. (2009). Protochlorophyllide: a new photosensitizer for the photodynamic inactivation of Gram–positive and Gram–negative bacteria. *FEMS Microbiol. Lett.*, *290*, 156–163.

Wang, Q., Hall, C. L., Al–Adami, M. Z. & He, Q. (2010). IsiA is required for the formation of photosystem I supercomplexes and for efficient state transition in *Synechocystis* PCC 6803. *PloS ONE*, *5*(5), e10432, 1–10.

Wang, Q., Jantaro, S., Lu, B., Majeed, W., Bailey, M. & He, Q. (2008). The high light–inducible polypeptides stabilize trimeric photosystem I complex under high light conditions in *Synechocystis* PCC 6803. *Plant Physiol.*, *147*, 1239–1250.

Warren–Rhodes, K. A., Rhodes, K. L., Pointing, S. B., Ewing, S. A., Lacap, D. C., Gómez–Silva, B., Amundson, R., Friedmann, E. I. & McKay, C. P. (2006). Hypolithic cyanobacteria, dry limit of photosynthesis, and microbial ecology in the hyperarid Atacama Desert. *Microb. Ecol.*, *52*, 389–398.

Waterbury, J. B., Watson, S. W., Guillard, R. R. L. & Brand, L. E. (1979). Widespread occurrence of a unicellular, marine, planktonic, cyanobacterium. *Nature*, *277*, 293–294.

Wierzchos, J., Ascaso, C. & McKay, C. P. (2006). Endolithic cyanobacteria in halite rocks from the hyperarid core of the Atacama Desert. *Astrobiology*, *6*, 415–422.

Wilkinson, F., Helman, W. P. & Ross, A. B. (1995). Rate constants for the decay and reactions of the lowest electronically excited singlet state of molecular oxygen in solution. An expanded and revised compilation. *J. Phys. Chem. Ref. Data*, *24*(2), 663–1021.

Willows, R. D. (2003). Biosynthesis of chlorophylls from protoporphyrin IX. *Nat. Prod. Rep.*, *20*, 327–341.

Wilson, A., Ajlani, G., Verbavatz, J.–M., Vass, I., Kerfeld, C. A. & Kirilovsky, D. (2006). A soluble carotenoid protein involved in phycobilisome-related energy dissipation in cyanobacteria. *Plant Cell*, *18*, 992–1007.

Wilson, A., Boulay, C., Wilde, A., Kerfeld, C. A. & Kirilovsky, D. (2007). Light–induced energy dissipation in iron–starved cyanobacteria: roles of OCP and IsiA proteins. *Plant Cell*, *19*, 656–672.

Wilson, A., Kinney, J. N., Zwart, P. H., Punginelli, C., D'Haene, S., Perreau, F., Klein, M. G., Kirilovsky, D. & Kerfeld, C. A. (2010). Structural determinants underlying photoprotection in the photoactive orange carotenoid protein of cyanobacteria. *J. Biol. Chem.*, *285*(24), 18364–18375.

Winterbourn, C. C. & Metodiewa, D. (1994). The reaction of superoxide with reduced glutathione. *Arch. Biochem. Biophys.*, *314*, 284–290.

Wu, Y. P. & Krogmann, D. W. (1997). The orange carotenoid protein of *Synechocystis* PCC 6803. *Biochim. Biophys. Acta*, *1322*, 1–7.

Yao, D., Kieselbach, T., Komenda, J., Promnares, K., Hernández–Prieto, M. A., Tichy, M., Vermaas, W. & Funk, C. (2007). Localization of the small cab–like proteins in Photosystem II. *J. Biol. Chem.*, *282*(1), 267–276.

Zeeshan, M. & Prasad, S. M. (2009). Differential response of growth, photosynthesis, antioxidant enzymes and lipid peroxidation to UV–B radiation in three cyanobacteria. *S. Afric. J. Botany*, *75*, 466–474.

Zhang, H., Liu, H., Niedzwiedzki, D. M., Prado, M., Jiang, J., Gross, M. L. & Blankenship, R. E. (2014). Molecular mechanism of photoactivation and structural location of the cyanobacterial orange carotenoid protein. *Biochemistry*, *53*, 13–19.

Zhang, S.–P., Xie, J., Zhang, J.–P., Zhao, J.–Q. & Jiang, L.–J. (1999a). Electron spin resonance studies on photosensitized formation of hydroxyl radical by C–phycocyanin from *Spirulina platensis*. *Biochim. Biophys. Acta*, *1426*, 205–211.

Zhang, S.–P., Xie, J., Zhang, J.–P., Zhao, J.–Q. & Jiang, L.–J. (1999b). Characterization of the transient species generated by the photoexcitation of C–phycocyanin from *Spirulina platensis*: a laser photolysis and pulse radiolysis study. *Biochim. Biophys. Acta*, *1472*, 270–278.

Zhao, C., Yang, Y. & Zhang, Z. (2012). Photocatalytic treatment of microcystin–LR–containing wastewater using Pt/WO_3 nanoparticles under simulated solar light. *Open J. Appl. Sci.*, *2*, 86–92.

In: Cyanobacteria
Editor: Douglas Davison

ISBN: 978-1-61470-925-1

Chapter 2

The Involvement of Cyanobacteria in Petroleum Hydrocarbons Degradation: Fundamentals, Applications and Perspectives

Ioan I. Ardelean
Department of Microbiology, Institute of Biology Bucharest,
Romanian Academy, Romania

Abstract

This review focuses on cyanobacteria which are copiously studied in the last two decades with respect to their presence and activity in petroleum hydrocarbon or its derivates (PH) polluted environments. The ability of consortia of (selected) cyanobacteria and hydrocarbonoclastic heterotrophic bacteria (HCHB) to degrade PH is clearly demonstrated by an increasing number of scientific reports that strongly argue that these mixed populations can be very important participants in the bioremediation of PH polluted sites. There are clear results which show that HCHB work better together with cyanobacteria for the bioremediation of PH contaminated sites, the general view being that the degradation of PH is carried out predominantly by HCHB, the contribution of cyanobacteria being mostly indirect. This is an important contribution mainly :i) by providing the molecular oxygen needed for aerobic respiration of heterotrophic bacteria during oxic degradation of PH; ii) by producing and excreting organic substances which sustain the activity of HCHB and iii) by offering HCHB surfaces to attach to. Some future prospects are also presented, mostly form a microbiological perspective, concerning the strategies to enhance petroleum hydrocarbon degradation by increasing the basic knowledge about the biology of cyanobacteria at molecular, cellular/population and ecological level, and integration with other cyanobacteria-based biotechnologies.

Keywords: Cyanobacteria, hydrocarbonoclastic heterotrophic bacteria, petroleum hydrocarbons degradation, bioremediation

1. INTRODUCTION

Cyanobacteria are the most diversified, ecologically most successful and evolutionary most important group of prokaryotes, clearly defined by their ability to carry out oxygenic photosynthesis in the thylakoide membranes and respiration both in plasma membrane and thylakoide membrane; in some species the major respiratory activity is occurring in the plasma membrane whereas in other species the main site of aerobic respiration is the thylakoide membrane (Peschek, 1996). Historically, cyanobacteria were the first organisms to perform oxygenic photosynthesis and, because of this metabolic ability of ancient cyanobacteria, the molecular oxygen started to accumulate 3.5 billion years ago in the reducing atmosphere of Earth. Cyanobacteria, being very versatile microorganisms, can live free or attached, as they are or in symbiosis with different types of organisms, in very different environments with respect to temperature, pH, salinity, light intensity, etc., some of them being able to use atmospheric nitrogen for the synthesis of organic compounds (for more information on cyanobacteria see e.g. Bryant 1994; Seckbach, 2007; Gault and Marler 2009; Pescheck et al. 2011; Govindgee and Shevela, 2011; Whitton 2012; Srivastava et al. 2013). Furthermore, some cyanobacteria are able to synthesize diverse hydrocarbon secondary products and by their biomass seem to be involved in the biogenesis of oil (see Radwan and Al-Hassan, 2000 and Al-Thani and Potts, 2012 for more details).

Crude oil or petroleum refers to natural deposits of complex mixtures of different types of hydrocarbons belonging to the following four classes (Speight, 1999): naphthenes, paraffins, aromatics and asphaltics. The proportion of each class of these hydrocarbons varies from one deposit to another, and determines the physical and chemical properties of that specific oil. The use of petroleum generates pollution by oil spills; major oil spills occurred following tanker naufrages, war actions, and natural reservoir leacking. The contribution of heterotrophic bacteria to PH degradation is very well documented (e.g. Leahy and Colwell, 1990; Head and Swannel, 1999; Cravo-Laureau and Duran, 2014; Fatherpure, 2014). The involvement of cyanobacteria in natural attenuation of crude oil pollution became clearly visible only after the Gulf War (August 2, 1990-February 26, 1991); on 25th of January 1991 a huge quantity of oil (almost 1.5 million tons) were deliberately released from a Kuwaiti terminal into the Arabian Gulf, covering around 1.500 km^2 of sea surface, severely polluted stretches of about 770 km (Radwan and Al-Hassan, 2000) out of which 480 km were truly blackened (Al-Thani and Potts, 2012). Toward the end of 1991 blue green mats, associated with oil, became macroscopically visible along the coast; the blue green mats were dominated by filamentous cyanobacteria (*Microleus chtenoplastes* and *Phormidium corium*) closely associated with HCHB belonging to genera *Rhodococcus*, *Arthrobacter*, *Pseudomonas* and *Bacillus* (Sorkhoh et al. 1992). The occurrence of these microbial mats dominated by cyanobacteria pushed scientists to open a new era in the study of the interaction between cyanobacteria and oil.

The aim of this chapter is to focus on: i) the involvement (indirect or/and direct) of cyanobacteria in the oxidation/ degradation of PH; ii) the interplay between cyanobacteria and HCHB in the presence of PH, as a particular case of the interaction between cyanobacteria with other microorganisms and iii) suggestions for some future prospects, mainly form a microbiological perspective.

2. Involvement of Cyanobacteria in the Oxidation/ Degradation of PH

Cyanobacteria are very versatile with respect to the metabolism of carbon (Raven, 2012) being photolithotrophs (light as source of energy and inorganic carbon as the only source of carbon), photo-organotrophs (light as source of energy and organic carbon as the source of carbon) and chemo-organotrophs (energy and carbon from organic carbon). Their ability to use organic substances as source of carbon was predominantly investigated in regard to glucides, organic acids as acetate, etc. (see Radwan and Al-Hassan, 2000, Al-Thani and Potts, 2012 and Raven, 2012 and references herein for more details). However, the relationship of cyanobacteria with PH received till 1992 far less attention than the ability to use carbon sources such as glucides and organic acids.

There are reports concerning the inhibitory effects of PH, or different fractions, or pure substances against cyanobacteria, with special emphasis on aerobic respiration, oxygenic photosynthesis and growth, which were clearly and amply reviewed by Radwan and Al-Hassan (2000). The type of effects seems to be bactericidal, but, up to my best knowledge, no detailed experiments have been done on the possible bacteriostatic/ cyanostatic activity of PH, or different fractions, or pure substances. In my opinion, a more detailed study of these oil-sensitive cyanobacteria could help to better understand the molecular target(s) and the mechanisms involved in the inhibition of cyanobacteria catabolism (e.g. respiration) and anabolism (photosynthesis, cell growth and division, etc.).

There are also reports on the oxidation of phenantrene by cyanobacteria (Cerniglia et al. 1980a, b) as well as on the stimulation of oxygenic photosynthesis, aerobic respiration and growth of cyanobacteria in the presence of low concentrations of n-alkanes (Gaur and Singh, 1990)

The ability of cyanobacteria to grow in the presence of petroleum hydrocarbons became evident when, as a consequence of the Gulf war, in the petroleum contaminated area microbial mats dominated by cyanobacteria become clearly visible (Sorkokh et al. 1992; Abed et al. 2002, 2006; Al- Hassan et al. 2001; Radwan and Al-Hassan, 2001a,b, Radwan et al. 2000, 2002, 2005, 2010; Sorkokh et al. 2010a,b; Al-Mailem et al. 2014; Al-Obaidy et al. 2014).

In this context, Al Hasan et al. 1994 showed experimental evidence that non axenic strains of two cyanobacteria, *Microleus chtneoplastes* and *Phormidium corium*, increase the proportion of their unsaturated fatty acids when grown in the presence of n-alkanes, thus behaving as typical hydrocarbon-utilizing heterotrophic bacteria. The growth of these cyanobacteria in the presence of crude oil or n-alkanes led to increased proportions of total unsaturated C16 and C18 fatty acids, suggesting the possibility that these cyanobacteria might be involved in n-alkane oxidation to fatty acids, as is the case in HCHB (Al Hasan et al. 1994). To follow this idea, the same scientific group designed further experiments showing that cyanobacterial biomass incubated in the presence of C14 and C16 chain alkanes contain fatty acids with the equivalent chain lengths in their lipids, as is the case of lipids specific to thylakoides, monogalactosyldiacylglycerols, digalagctosyldiacylglycerols and sulfoquinovossylidiacyl glycerols (Al Hassan et al. 1998). The authors stress on the fact that the results of all these experiments argue for the ability of these cyanobacteria to oxidize n-alkanes (see Radwan and Al-Hassan, 2000 for more details).

Kuritz and Wolk (1995) reported that two filamentous cyanobacteria, *Anabaena* sp. and *Nostoc* sp. are able to oxidize chlorinated cyclic aliphatic hydrocarbons and have the potential to degrade the pesticide lindane (δ-hexachlorocyclohexane), presenting quantitative evidence that this ability can be enhanced by genetic engineering: the addition of the *linA* gene increased the degradation of lindane by the two cyanobacteria strains, at least when they were grown with N_2. The paper further provides qualitative evidence that those two strains can be genetically engineered to degrade another chlorinated pollutant, 4-chlorobenzoate. The researches of Kuritz and Wolk (1995), are considered a pioneering work with respect to genetic engineering of cyanobacteria for PH degradation.

Safonova et al. (1999) compared the growth of six genera of Cyanobacteria (*Phormidium, Lyngbya, Oscillatoria, Nostoc, Anabaena, Microcystis*), in pure and in mixed cultures, with a standard oil-degrading preparation of *Rhodococcus* sp. 7HX. They found that the presence of alcanotrophic bacteria restores the growth of oil-sensitive cyanobacteria strains and stimulates growth in tolerant strains. Some strains (*Phormidium* sp. ES- 90, O-b-1; *Nostoc* sp. U-15:3) in combination with alcanotrophic bacteria, demonstrated very high resistance to black oil pollution; furthermore, their association with alcanotrophic bacteria destroyed black oil more efficiently than the pure alcanotrophic bacterium *Rhodococcus* sp. 7HX culture (Safonova et al. 1999).

Raghukumar et al. 2001 showed that the marine cyanobacteria *Oscillatoria salina, Plectonema terebrans* and *Aphanocapsa* sp. degraded Bombay High crude oil when grown in artificial seawater as well as in plain natural seawater: around 45-55% of the total fractions of crude oil (containing 50% aliphatics, 31% waxes and bitumen, 14% aromatics and 5% polar compounds) were removed in the presence of these cultures within 10 days. Furtmermore, mixed cultures of the three cyanobacterial species removed over 40% of the crude oil. Experiments done on pure hydrocarbons showed that between 50% and 65% of pure hexadecane (model aliphatic compound) and 20% to 90% of aromatic compounds (anthracene and phenanthrene) disappeared within 10 days.

The involvement of unicellular cyanobacteria in the fate of PH in marine environment was also investigated. Epifluorescent microscopic counting of picocyanobacteria (Al-Hassan et al. 2001) showed rich populations along the Kuwaiti coast of the Arabian Gulf, with cell densities of $1x10^5$ - $6x10^5$ per mL, most dominant being the *Synechococcus* genus and less dominant the *Synechocystis, Pleurocapsa and Dermocarpella* genera. Interestingly, their isolates can grow in inorganic medium supplemented with crude oil up to 0.1% (w/v). The analysis showed that representative strains of these four genera can accumulate in their inter-thylakoide spaces n-hexadecane and phenanthrene which can not be oxidized (Al-Hasan et al. 2001). The authors conclude that the associated heterotrophic bacteria are possibly involved in the biodegradation of the PH.

Cohen (2002) studied the response of two type of cyanobacterial mats towards crude oil; the pristine mat (Solar Lake) and the polluted one (African coast of Suez) showed efficient crude oil degradation in the light, followed by exuberant growth of *Phormidium* and *Oscillatoria*; however, the isolated axenic cyanobacteria strains are not able to degrade crude oil, the authors concluding that in natural conditions the degradation was done by heterotrophic contaminants, such as *Marinobacter* sp. Interestingly, in axenic cultures the exposure to crude oil is followed by a marked inhibition of photosynthesis (Cohen, 2002) which could suggest that, in microbial mats, the inhibition of photosynthesis, at shorter or longer time scale, is not so marked as in pure cultures.

Al Awadi et al. (2003) developed artificial microbial consortia of phototrophic microorganisms (cyanobacteria, picoplancton, diatoms) as primary colonizers of gravel particles (2-3 cm diameter) and glass plates, and HCHB (*Acinetobacter calcoaceticus* and *nocardiforms*) attached to the cyanobacterial filaments. These artificial microbial consortia developed as biofilms showed hydrocarbon removal capacity when potassium nitrate was added as fertilizer.

Safonova et al. (2004) selected several cyanobacteria-algal-bacterial consortia and selected the cyanobacterium *Phormidium* sp. ES-90 and different HCHB (*Rhodococcus* sp. Ac-1267, *Kibdelosporangium aridum* 754 as well as other two unidentified bacterial strains, St-1 and St-2) with good results concerning the reduction of the biological oxygen demand (BOD_{25}) and the chemical oxygen demand (COD) amounted to 97 % and 51 %, respectively.

Bleijswijk and Muyzer (2004) studied the genetic diversity of oxygenic phototrophs in microbial mats exposed to different levels of oil pollution. They found that cyanobacteria related to members of the *Phormidium* and *Planktotrix* genera were present in the polluted site, indicating PH tolerance; however, in heavily polluted mats, only *Oscillatoria*-related cyanobacteria could persist.

De Oteyza et al. (2004) showed that *Microcoleus* consortia can grow in the presence of two different types of crude oil, one with high content in aliphatic hydrocarbons (Casablanca) and the other rich in sulphur and aromatic compounds (Maya). Besides the physical contact between cyanobacteria and the crude oil layer, small changes in oil composition were obtained only for aromatic-rich oil, the other components remaining unchanged.

Sánchez et al. (2005) performed a study in cultures of *Microcoleus* grown in mineral medium and in cultures of the cyanobacterium grown in mineral medium supplemented with crude oil. The results indicate that most of the clones found in the polluted culture correspond to well-known oil-degrading and nitrogen-fixing microorganisms, and belong to different phylogenetic groups, such as the Alpha, Beta, and Gamma subclasses of Proteobacteria, and the Cytophaga/Flavobacteria/Bacteroides group. The control is dominated by one microorganism (88% of the clones) closely affiliated to *Pseudoxanthomonas mexicana* (similarity of 99.8%). The presence of organisms closely related to well-known nitrogen fixers such as *Rhizobium* and *Agrobacterium* suggests that at least some of the cyanobacteria-associated heterotrophic bacteria are responsible for nitrogen fixation and degradation of hydrocarbon compounds inside the polysaccharide sheath, whereas *Microcoleus* provides a habitat and a source of oxygen and organic matter.

Abed and Koster (2005) investigated the potential of 10 mat-forming cyanobacteria from saline environments to degrade different types of hydrocarbons (phenanthrene, pristane, *n*-octadecane, and dibenzothiophene); interestingly, in order to enhance the physical contact between cyanobacteria and hydrocarbons, the authors used hydrophobic clay. Five strains (*Aphanothece halophyletica*, *Dactyolococcopsis salina*, *Halothece* strain EPUS, *Oscillatoria* strain OSC, and *Synechocystis* strain UNIGA) were able to degrade *n*-alkanes. In case of the other five strains (*Microcoleus chthonoplastes*, *Oscillatoria* sp. MPI 95 OS 01, *Halothece* strain EPUG, *Halomicronema exentricum*, and *Phormidium* strain UNITF) the alkanes were not significantly affected. They concluded that the cyanobacteria-associated aerobic heterotrophic bacteria were responsible for the observed biodegradation, cyanobacteria playing an indirect role in biodegradation by supporting the growth and activity of the actual degraders (Abed and Koster, 2005).

Chaillan et al. (2006) showed that a cyanobacterial mat from a petroleum-contaminated tropical environment is very efficient in degrading the crude oil both in the environment and in a laboratory microcosms model. The main component (biomass basis) of this cyanobacterial mat is the cyanobacterium *Phormidium animale*, other phototrophs and heterotrophic bacteria being also present. However, in axenic culture, *Phormidium animale* did not exhibit any degradative capacity on hydrocarbons in the range of C13–C35 carbon atom number either in autotrophic or heterotrophic conditions, the authors concluding that the degradation of crude oil is done exclusively by the other microorganisms present in the microbial consortium of the mat (Chaillan et al. 2005).

Other biodegradation experiments were conducted with locally selected consortia of microorganisms, the most abundant cyanobacterium being *Phormidium foveolarum* and the most abundant microalga being *Achanthes minutissima* (Antic et al. 2006); the previous grown biomass of phototrophs was added to crude Sirakovo (Sir, paraffinic type) and Velebit (Ve, naphthenic type) oils in the presence of either inorganic medium or organic medium. The results obtained after three months of experiments showed that the petroleum biodegradation was restricted to the acyclic aliphatics (n-alkanes and isoprenoids), in the inorganic medium in the light (Antic et al. 2006). These results seem to sustain the involvement of phototrophs in petroleum degradation, but no details are available with respect to either direct or indirect role of phototrophs in petroleum degradation, or to the quantification of heterotrophic bacteria contribution.

Lliròs et al. 2008 studied two aspects: i) the influence of microbial mats on the degradation of two crude oils (Casablanca - aliphatic with a low viscosity and Maya - sulphur-rich heavy crude oil that is predominantly aromatic) and ii) the effect of oil pollution on the mat structure. The results showed that in the unpolluted microcosms, *Microcoleus chthonoplastes*-, *Phormidium*- and *Oscillatoria*-like were the dominant filamentous cyanobacterial morphotypes, whilst *Synechoccocus*-, *Synechocystis*- and *Gloeocapsa*-like were the most abundant unicellular cyanobacteria. Among filamentous cyanobacteria, *M. chthonoplastes*-like morphotype was the most resistant to both oils, unlike the other cyanobacteria, which tolerated Casablanca but not Maya. Unicellular cyanobacteria seemed to be resistant to pollution with both used oils, with the exception of the morphotype resembling *Gloeocapsa*, which was sensitive to both oils.

Chavan and Mukherji (2008, 2010a,b), working on mixed consortia of cyanobacteria, microalgae and HCHB, claim that the indirect role of cyanobacteria in facilitating hydrocarbon biodegradation is done by: a) providing the hydrocarbon-degrading bacteria with molecular oxygen, fixed nitrogen, and vitamins; b) feeding oil-degrading heterotrophic bacteria, where no petroleum is available, with some products of their own photosynthesis which are secreted in the external medium; c) facilitating the immobilization of the hydrocarbon degraders as biofilms on different surfaces; d) cyanobacterial polysaccharides which play a major role in the emulsification of oil, i.e., breaking of oil into small droplets, which subsequently facilitate degradation by heterotrophs.

Ibraheem (2010) isolated five cyanobacterial species (*Phormidium* sp., *Nostoc* sp., *Anabaena* sp., *Aphanothece conferta,* and *Synechocystis aquatilis*) which were tested for biodegradation of four hydrocarbon compounds: two aliphatic compounds (*n*-octadecane and pristane) and two aromatic compounds (phenanthrene and dibenzothiophene). The best results with respect to high degradation efficiencies were obtained by *A. conferta* (64% for *n*-octadecane and 78% for pristane) and by *S. aquatilis* (85% for *n*-octadecane and 90% for pristane). Abed (2010) studied the interaction between cyanobacteria and aerobic heterotrophic bacteria in the degradation of hydrocarbons. In order to test 1) whether *Synechocystis* PCC6803 could grow on hexadecane, 2) whether the presence of associated aerobic heterotrophs influence the growth of the cyanobacterium and 3) whether the growth of the aerobic heterotrophs on hydrocarbons might stimulate/inhibit the growth of the cyanobacterium, the author (Abed, 2010) monitored the growth of (initially axenic) *Synechocystis* PCC6803 under the following conditions: 1) *Synechocystis* PCC6803 alone; 2) *Synechocystis* PCC6803 together with the GM41 strain in the presence of hexadecane (1% v/v); 3) *Synechocystis* PCC6803 in the presence of hexadecane (1% v/v) and 4) *Synechocystis* PCC6803 + the GM41 strain without hexadecane. The growth of *Synechocystis* PCC6803 (axenic culture) in the presence of hexadecane reaches similar values as the control (*Synechocystis* PCC6803 alone, axenic culture), showing that cyanobacterial growth is not affected (either positive or negative) by the presence of hexadecane (1% v/v). The growth of *Synechocystis* PCC6803 (axenic culture) together with the GM41 strain (but without hexadecane) stimulated the growth of the cyanobacterium, a maximum growth (measured as chlorophyll *a* concentration) being recorded after 20 days, suggesting the occurrence of beneficial effects on cyanobacterial growth. Interestingly, the best growth of *Synechocystis* PCC6803 strain was observed in the presence of hexadecane and the GM41 strain, indicating that hexadecane catabolisation by the GM41 strain sustains the growth of cyanobacterium. The growth of the cyanobacterium in the presence of both hexadecane and the GM41 strain was 2 times higher than with only the GM41 strain and 12 times higher than without both (Abed, 2010). When it comes to the effect of organic substances known to be cyanobacterial exudates (acetate, pyruvate, glucose, alanine and butanol) on phenantrene consumption by GM42 strain, the situation is diverse. The addition of alanine and butanol did not show any significant effect on phenanthrene degradation and the consumption rate was comparable to the control (0.34 μg/day). The addition of glucose, pyruvate and acetate significantly stimulated the degradation process, the consumption rate of phenantrene increasing to 0.48-0.54 μg/day. The author concludes that the cyanobacterium *Synechocysti*s PCC 6803, the most abundant phototrophic microorganism in the natural mats they discovered, grew better in the presence of the associated aerobic heterotrophic bacteria and could provide them with the necessary organics for efficient degradation activities, claiming that the aerobic heterotrophs-cyanobacterial mat consortia offer a cost-effective and promising system for bioremediation of oil-polluted coastal sites (Abed, 2010).

Al-Bader et al. (2012) showed that the air-dust-borne associations of phototrophic and oil vapor-utilizing bacteria are able to produce hydrocarbon attenuation in batch cultures; as there are no differences in hydrocarbon consumption in dark or in light, the authors logically concluded that attenuation activities were due to the heterotrophic bacteria. The contribution of phototrophic microorganisms (the two cyanobacteria *Nostoc commune* and *Leptolyngbya thermalis* and the chlorophyte (Domain *Eukarya*) associated with dust seems to be an , indirect one by : i) providing the heterotrophic bacteria with oxygen, a by-product of photosynthesis which is involved in the initial step of microbial attack on the hydrocarbon molecules (Klug and Markovetz, 1971; Rehm and Reiff, 1981) ; ii) efficiently keeping the water content in the air at an acceptable level as well as by iii) the enhancement of heterotrophic bacterial growth by unidentified water-soluble and diethyl ether soluble metabolites. The authors concluded that the airborne microbial associations may be effective in bioremediation of atmospheric hydrocarbon pollutants *in situ* (Al-Bader et al. 2012).

There are reports on the establishment of biofilms containing cyanobacteria and hydrocarbon-utilizing heterotrophic bacteria in laboratory in order to degrade petroleum hydrocarbons (Radwan and Al-Hasan 2001a). Furthermore the systems were improved by the inclusion of diazotrophic bacteria (Al-Bader et al. 2012) and deeper characterization by molecular methods (denaturing gradient gel electrophoresis (DGGE) of PCR-amplified rRNA gene fragments in the total DNA extracts from the biofilms, followed by band amplification and sequencing). The authors (Al-Bader et al. 2012) claim that laboratory-established phototrophic biofilms containing hydrocarbon-utilizing (heterotrophic) bacteria, able to fix atmospheric nitrogen, would be effective tools both for fundamental research and applications related to hydrocarbon bioremediation in aquatic environments.

Amirlatifi et al. (2013) reported that their isolate *Microchaete tenera* ISC13 grown in BG_0 and treated with different oil concentrations (1, 2.5, 5 and 7%) showed the following main results: i) cyanobacterium growth increased with elevated oil concentration, but no changes were observed in chlorophyll content whereas phycocyanin and allophycocyanin decreased in the presence of oil, having the highest rate in control. They do not report any effect on nitrogen fixation, but the occurrence and density of heterocysts as well as the growth of cyanobacterium suggest that the nitrogen fixation is not limiting the growth of this cyaobacterium in the presence of oil.

Babaei et al. (2013) showed that *Anabaena* sp. ISC55, isolated from an oil polluted area, treated with different crude oil concentrations (0; 0.5; 1; 2.5; 5, and 6%), exhibits interesting reactions: a) the cellular growth increased with higher concentrations of oil − except 6% oil; b) photosynthetic activity and chlorophyll *a* and phycobiliproteins content decreased with increasing oil concentration; c) nitrogenase activity was completely inhibited by adding crude oil to the cultures. The authors concluded that this strain seems to be a valuable strain for further studies as it can reduce oil content by 8 and 53% after 14 and 28 days, respectively (Babaei et al. 2013).

Chronopoulou et al. (2013) observed that oil pollution in a mudflat sediment mesocosm led to an increase in benthic diatoms and a large increase in cyanobacteria. The increased photoautotrophic growth, together with the activity of aerobic hydrocarbon degrading bacteria, resulted in nitrogen depletion, which most likely selected for dinitrogen-fixing cyanobacteria, eventually resulting in their co-dominance along with diatoms.

Zhubanova et al., (2013) reported the use of a special methodology to obtain bacteriological pure (axenic) cultures of cyanobacteria: *Phormidium* sp. K-1, *Oscillatoria* sp. A-2, and *Oscillatoria* sp. C-3 that were further used together with selected HCHB belonging to the *Pseudomonas, Alcaligenes, Arthrobacter*, and *Bacillus* genera to construct artificial cyanobacterial–bacterial communities with a high hydrocarbon-oxidizing activity. The best results in field experiments were obtained with cyanobacterial–bacterial associations between *Phormidium* sp. K-1 sp. and *Pseudomonas stutzeri* A1, *Pseudomonas* sp. N2, and *P. alcaligenes* A5. The authors concluded that the obtained results are the basis for the elaboration of microbiological technologies for environment protection using cyanobacteria (Zhubanova et al. 2013).

Cole et al. (2014), in a very complex approach, examined the primary succession in phototrophic communities, namely two unicyanobacterial consortia isolated and enriched from a microbial mat in Hot Lake, Washington. The authors identified the strains, and higher taxonomic groups, characterizing the membership by molecular techniques and metabolic function of each consortium by metabolome analysis. Furthermore, they analyzed the dynamic of spatial structures/ biofilm and the density of ADN containing bacterial cells. At higher taxonomic levels, the two consortia exhibited similar assembly patterns, with increasing relative abundances of members from Bacteroidetes and Alphaproteobacteria concurrent with decreasing relative abundances of those from Gammaproteobacteria. Interestingly, when it comes to species distribution within these groups, the relative abundances of individual heterotrophic species were substantially different in the developing biofilms. This suggests that, although similar niches are created by the cyanobacterial metabolisms, the resulting webs of autotroph-heterotroph and heterotroph-heterotroph interactions are specific to each primary producer (Cole et al. 2014).

3. Relationships between Cyanobacteria and HCHB in the Presence of Petroleum Hydrocarbons

Cyanobacteria interact with other microorganisms both in the absence (Safferman and Morris, 1962; Bershova et al. 1968; Bryant 1994; Seckbach, 2007; Gault and Marler 2009; Pescheck et al. 2011; Govindgee and Shevela, 2011; Whitton 2012; Srivastava et al. 2013) of PH and in its presence (see McGenity et al. 2012 and references herein). I will try here to classify these interactions only between cyanobacteria and HCHB.

Table 1 presents all the theoretical possible relationships taking into account the direct effect of PH-tolerant/ oxidant cyanobacteria upon HCHB, indicating when there is experimental evidence for a given type of interaction.

Table 1. Theoretical relationships between PH-tolerant/ oxidant cyanobacteria and HCHB (for more explanations see the text), taking into account the direct effect between species/populations

Cyanobacteria	HCHB	NAME	DEFINITION	KOWN SO FAR
0	+	Commensalism	HCHB takes some advantages whereas cyanobacteria are not affected	YES
0	0	Neutralism	No interactions	No
0	-	Amensalism	HCHB inhibited by cyanobacteria	No
-	+	Parasitism/ Predation	HCHB feed on living cyanobacteria	No
-	0	Amensalism	Cyanobacteria inhibited by HCHB	No
-	-	Competition	Competition for energy, nutrients, space etc.	No
+	+	Mutualism/ protocooperation	Both cyanobacteria and HCHB benefit	YES
+	**0**	Commensalism	Cyanobacteria take some advantages whereas HCHB are not affected	YES
+	-	Parasitism/ Predation	Cyanobacteria feed on living HCHB	No

As can be seen in table 1, there are several types of direct interactions between PH-tolerant/ oxidant cyanobacteria and HCHB, which are documented in the literature: commensalism, mutualism and protocooperation.

Commensalism is a type of relationship between two populations or species in which one of them benefits from the association whereas the other one is not affected at all.

In the case when cyanobacteria benefit, there are some classical papers showing the positive effect of carbon dioxide produced by heterotrophic bacteria on cyanobacterial metabolism. In a study carried out on 12 strains of cyanobacteria grown in association with 22 strains of heterotrophic bacteria able to use as carbon sources different types of aliphatic hydrocarbons, Lange (1971) showed a stimulation of cyanobacterial growth, probably as a result of carbon dioxide supplied by heterotrophic bacteria during their aerobic respiration (Lange, 1971). This explanation is sustained also by the fact that, when the atmosphere was enriched with 0.5% CO_2, an increase in cyanobacterial growth was recorded as well, but there was no evidence of additional, supplementary aliphatic hydrocarbon metabolisation by heterotrophic bacteria. Interestingly, axenic cultures of these cyanobacteria are not able to use aliphatic hydrocarbons (Lange, 1971). These results suggest that the respiratory carbon dioxide produced by heterotrophs is used as raw material for autotrophic growth of the photosynthetic bacteria, the increasing of respiratory carbon dioxide level being one mechanism by which heterotrophic bacteria can enhance the growth of cyanobacteria. Other examples of commensalism with benefits for cyanobacteria are those in which HCHB are well adapted to sequestering iron or to decrease molecular oxygen or PH concentration in the vicinity of cyanobacterial cells, filaments or aggregates (more details see McGenity et al. 2012).

One case of commensalism, when HCHB benefit from the presence and activity of cyanobacteria, concerns the photosynthetic production of molecular oxygen which is needed for the aerobic degradation of PH by HCHB. Many cyanobacteria are able to produce different types of hydrocarbons that could be used as (an alternative) carbon and energy source by HCHB (for more details see McGenity et al. 2012). The ability of some cyanobacteria to assimilate atmospheric nitrogen in nitrogen-limited pristine and PH polluted marine sediments reconstructed in aquaria (Musat et al. 2006) should be an important contribution to nitrogen cycle at that site, providing microbiota, including HCHB, with an essential nutrient for their life (Knowles and Wishart, 1977; Thomson and Webb, K, 1984; Toccalino et al. 1993). Furthermore, recent results (Morales and Paniagua-Michel, 2013) show that surfactants produced by a *Phormidium* strain contribute to hexadecane and diesel oil degradation by laboratory- developed biofilms with this cyanobacterium.

The ability of cyanobacteria to produce exopolysaccharides and to develop stratified structures such as biofilms and mats in oil–polluted ecosystems, the largest example being the cyanobacteria-dominated mats in Arabic Golf (Sorkokh, et al. 1992; Abed et al. 2002, 2006; Al- Hassan et al. 2001; Radwan and Al-Hassan, 2001a,b; Radwan et al. 2000, 2002, 2005, 2010; Sorkokh et al. 2010a,b; Al-Mailem et al. 2014; Al-Obaidy et al. 2014), argue for very complex interactions between cyanobacteria and HCHB which could reach the level of protocooperation or mutualism. Protocooperation means that both strains/ populations benefit from their relationship, but the interplay is facultative, each strain being able to live also in the absence of the other strain. In the case of mutualism the interaction is obligatory for both strains/ populations. The scientists claim that the tridimensional structure of the mats, in which cyanobacteria produce exopolysaccharides and further provide by their filaments a space for HCHB to attach to, are very important for degradation of PH which is carried out mainly by HCHB. The emergence, occurrence and persistence of cyanobacteria and heterotrophic associations as microbial mats in the presence of petroleum hydrocarbons (Arabic Golf) logically argue that the level of their interplay should cover the concept/ notion of mutualism or protocooperation. Even in the situation when (filamentous) cyanobacteria only offer room for HCHB to accumulate at higher densities, and to remain within these macroscopic structures (cyanobacterial biomass) as complex ecosystems, either as biofilms or as mats, the relationship is protocooperation. Further research will add more precise knowledge.

CONCLUSION

Cyanobacteria the most diversified, ecologically most successful, and evolutionary most important group of prokaryotes, are copiously studied in the last two decades with respect to their presence and activity in PH polluted environments, together with HCHB and other microorganisms. The ability of consortia of (selected) cyanobacteria and HCHB to degrade PH is clearly demonstrated by an increasing number of scientific reports that strongly argue that these mixed populations can be very important in the bioremediation of PH polluted sites. There are obvious results which show that HCHB work better in the presence of cyanobacteria for the bioremediation of PH polluted sites, the general view being that the degradation of PH is carried out mainly by HCHB, the contribution of cyanobacteria being mostly indirect, but very important, especially by providing molecular oxygen needed for the oxic degradation of PH, by producing and excreting organic substances which sustain the activity of HCHB and by offering space for the adhesion of HCHB to the surface of their cells and filaments, assemblies of cells and filaments, mainly when they live in stratified (micro)ecosystems (biofilms and mats), previously known as very robust structures (Paerl et al. 2000).

Most of cyanobacteria strains are sensitive to PH, being unable of cellular growth and multiplication in the presence of PH, whereas the strains found in different polluted sites, and further enriched by selective cultivation (together with heterotrophic bacteria, some of them being HCHB) are at least PH-tolerant cyanobacteria; the ability, if any, of these tolerant cyanobacteria to degrade PH deserves further attention. The mechanisms of oil tolerance in cyanobacteria is far from being understood, although abundant research has been published on these mechanisms in other Gram-negative bacteria, heterotrophic ones, and in Gram-positive bacteria.

The ability of our selected cyanobacterial populations to produce biosurfactants was showed by oil spread method and by emulsification assay (Ramos et al. 2002; Safonova et al. 2004; Lăzăroaie, 2010; Murínová and Dercová, 2014). In figure 1 one can see the macroscopic aspect of cyanobacterial populations enriched in the presence of PH, the microscopic pictures based on their natural fluorescence (red filter) and their ability to produce biosurfactants.

Figure 1. a) Macroscopic view of selected cyanobacterial populations able to grow in the presence of PH, b) and their natural fluorescence as seen by epifluorescence microscopy (red filter); the ability to produce biosurfactants in shown by oil spread method : c) oil film without cell culture and d) oil film disrupted by 10 μL cell culture; e)PH emulsification by cyanobacterial culture(e2) as compared with control (e1- PH and sterile culture medium).

The complex study of the biology of these tolerant cyanobacteria, following the work done on HCHB (see below) will show if naturally occurring cyanobacteria comprise some strains which, by their own, can degrade PH. Taking into account the results obtained so far in this field, I expect that the increasing research on the use of cyanobacteria and HCHB for bioremediation of PH polluted sites will become a viable commercial biotechnology as, for example, mass cultivation of *Arthrospira* is nowadays.

FUTURE PROSPECTS

In my opinion the following directions would improve our ability to use cyanobacteria in either pure cultures or co-cultures with HCHB to degrade different types of PH in natural ecosystem, in engineered ecosystems, or bioreactors:

1) In order to have a knowledge-based selection of cyanobacteria able to tolerate or degrade PH, the cyanobacteriologists should study cyanobacteria (creatively) following the work already done in HCHB concerning the changes occurring in the presence of PH. It is well known for HCHB that these changes concern the chemical composition of the external membrane, peptidoglycan layers and plasma membrane, efflux pumps and solvent transport, enzymatic and metabolic reactions (for more details see Ramos et al. 2002; Safonova et al. 2004; Lăzăroaie, 2010; Murínová and Dercová, 2014) as well as positive taxies towards PH (Marx and Aitiken, 2000, Lacal et al. 2011, Krell et al. 2013). Natural (enriched) cyanobacterial populations or/and already empirically selected PH-tolerant/ oxidant cyanobacteria strains (Ghita et al. 2013; Cao et al. 2013), including the (poly)extremophilic ones, should be studied at molecular, cellular/ individual and population level with respect to their structure and functions, including tolerance and degradation of PH in axenic cultures and co-cultures with HCHB.
2) A deeper understanding of intermediary metabolism in cyanobacteria in the presence of PH and HCHB would help us to see the differences, if any, in the ability of cyanobacteria with TCA-complete and TCA- incomplete cycles to contribute, either directly or indirectly, to PH degradation.
3) More attention should be payed to the possibility that cyanobacteria could use PH not only in catabolic reactions, but also as building blocks in anabolic reactions, as shown for the first time by Al Hassan et al. (1998) who suggested that cyanobacteria might be involved in n-alkane oxidation of corresponding fatty acids.
4) Following the studies concerning the effect of grazers on PH degradation by HCHB (Safonova et al., 1999; Sauret et al. 2012; Beaudoin et al., 2014), similar research should be done on PH degradation by cyanobacterial population, alone or in co-cultures with HCHB. The precise role of grazers in PH degradation by co-cultures of cyanobacteria and HCHB, either in nutrient rich or nutrient limited environments, would help to design engineered consortia with maximized ability to degrade PH by promoting or not the presence and activity of different types of grazers.
5) A better understanding of the interplay between respiration and photosynthesis, under so-called normal conditions, in both light and darkness (Ardelean, 2006;

Ardelean and Pescheck, 2011), but also during PH degradation by pure cultures of cyanobacteria or by co-cultures of cyanobacteria and HCHB (in the presence or in the absence of different types of grazers).

6) The effect of different dispersants on cyanobacteria with special emphasis on PH tolerance/degradation by cyanobacteria, either in pure or co-cultures with HCHB.
7) To deeper understand the interaction between cyanobacteria and other microorganisms in oil polluted ecosystems, with special emphasis on HCHB as well as tolerant phototrophic and heterotrophic microorganisms.
8) Comparative studies on PH tolerance/degradation of (a) given strain(s) of cyanobacteria, in pure and co-cultures, in different physical states: planktonic, benthic, (young) biofilm (including bacterioneuston or surface mono layer) or stratified ecosystem- mats, stromatolites or mature (multilayer) biofilms.
9) The use of (very) oil-sensitive cyanobacteria (whose metabolic activity mainly respiration and/or photosynthesis should be rapidly and drastically inhibited by very small amounts of PH) as probes/ bioreporters for PH or some of its components; the development of a network of (cheap) biosensors/bioreporters with both (very) oil-sensitive cyanobacteria and oil-degrading cyanobacteria (whose metabolic activity mainly respiration and/or photosynthesis should be rapidly and drastically sustained/ enhanced by very small amounts of PH) for rapid detection/ signaling of the presence of PH in different environments.
10) The knowledge concerning PH synthesis by cyanobacteria could enhance our abilities to use cyanobacteria for the degradation of PH. For example, in a recent review, Coates et al. 2014 showed that hydrocarbon biosynthetic pathways were identified in one-hundred-forty-two strains of cyanobacteria over a broad phylogenetic range. Based on literature reports and their own results, the authors indicate that hydrocarbon production is a universal phenomenon in cyanobacteria, all cyanobacteria possessing the capacity to produce hydrocarbons from fatty acids, using three different biochemical pathways (Coates et al. 2014).
11) Better understanding the interplay between cyanobacteria (including *Procholorococcus* chl b- containing phototrophic prokaryote) and oxygenic photosynthetic eukaryotes in PH degradation.
12) Genetic and metabolic engineering of oil-tolerant and oil-oxidant cyanobacteria with respect to catabolic and anabolic pathways, in order to promote their ability to degrade PH either in pure culture or in co-culture with HCHB. This could be done considering the progresses already achieved in the genetic and metabolic engineering of cyanobacteria for other biotechnological applications (Hernández-Prieto et al., 2014).
13) The coupling, if possible, of PH degradation with other potential relevant applications of cyanobacteria such as: biomass and related product synthesis, carbon dioxide mitigation, metal nanoparticule synthesis. etc., and assessing their ecological effects.
14) Microcosms (indoor and outdoor) as well as mesocosms experiments concerning PH degradation by cyanobacteria with respect to biostimulation (nutrient addition) and bioaugmentation (addition of previously grown cyanobacterial biomass), with or without dispersant(s) addition.

ACKNOWLEDGMENTS

This paper is dedicated to the memory of my respected master, Professor Gheorghe Zarnea (September 22, 1921 –June 16, 2012), from the University of Bucharest and the Institute of Biology Bucharest (Romania). His distinction and intelligence impressed and guided his numerous students, including me, to whom his enthusiasm was lastingly transmitted.

This work was supported by the Romanian Academy (Grant RO1567-IBB05/2014). Thanks are due to biologist Mihaela Mazanet for English corrections.

REFERENCES

Abed, R.M.M. (2010) Interaction between cyanobacteria and aerobic heterotrophic bacteria in the degradation of hydrocarbons. *Int.Biodeterior. Biodegrad. 64*, 58–64.

Abed, R.M.M., Safi, N.M.D., Koster, J., de Beer, D., El-Nahhal, Y., Rullkotter, J. & Garcia-Pichel, F. (2002) Microbial diversity of a heavily polluted microbial mat and its community changes following degradation of petroleum compounds. *Appl Environ Microbiol. 68*, 1674–1683.

Abed, R.M.M. & Köster, J. (2005) The direct role of aerobic heterotrophic bacteria associated with cyanobacteria in the degradation of oil compounds. *Int Biodeterior Biodegrad. 55*, 29–37.

Abed, R.M.M., Al-Thukair, A. & de Beer, D. (2006) Bacterial diversity of a cyanobacterial mat degrading petroleum compounds at elevated salinities and temperatures. *FEMS Microbiol Ecol 57*, 290–301.

Al-Awadhi, H., Al-Hasan, R.H., Sorkhoh, N.A., Salamah, S. & Radwan, S.S. (2003) Establishing oil-degrading biofilms on gravel particles and glass plates. *Int Biodeter Biodeg. 51*, 181–185.

Al-Awadhi, H., Al-Mailem, D., Dashti, N., Khanafer, M. & Radwan, S.S. (2012) Indigenous hydrocarbon-utilizing bacterioflora in oil-polluted habitats in Kuwait, two decades after the greatest man-made oil spill. *Arch Microbiol. 194*, 689–705.

Al-Bader, D., Eliyas, M., Rayan, R. & Radwan, S.S. (2012) Air–dust-borne associations of phototrophic and hydrocarbon-utilizing microorganisms: promising consortia in volatile hydrocarbon bioremediation. *Environ Sci Pollut Res. 19*, 3997–4005.

Al-Hasan, RH, Al-Bader, DA, Sorkhoh, NA, Radwan, SS (1998) Evidence for n-alkane consumption and oxidation by filamentous cyanobacteria from oil-contaminated coasts of the Arabian Gulf. Mar Biol 130: 521–527

Al-Hasan, R.H., Khanafer, M., Eliyas, M. & Radwan, S.S. (2001) Hydrocarbon accumulation by picocyanobacteria from the Arabian Gulf. *J Appl Microbiol. 91*, 533–540.

Al-Hasan, R.H., Sorkhoh, N.A., Al-Bader, D. & Radwan, S.S. (1994) Utilization of hydrocarbons by cyanobacteria from microbial mats on oily coasts of the Gulf. *Appl Microbiol Biotechnol. 41*, 615–619.

Al-Hasan, R.H., Khanafer, M., Eliyas, M. & Radwan, S.S. (2001) Hydrocarbon accumulation by picocyanobacteria from the Arabian Gulf. *J Appl Microbiol. 91*, 533–540.

Al-Mailem, D.M., Eliyas, M. & Radwan, S. (2014) Enhanced bioremediation of oil-polluted, hypersaline, coastal areas in Kuwait via vitamin-fertilization. *Environ Sci Pollut Res Int.* Mar;21, 3386-3394.

Al-Obaidy, A.H.M.J. & Lami, M.H.M. (2014) The Toxic Effects of Crude Oil in Some Freshwater Cyanobacteria. *Journal of Environmental Protection* 5, 359-367.

Al-Thani R.F., and Potts M., Cyanobacteria, Oil –and Cyanofuel? In Whitton B.A., editor.*Ecology of cyanobacteria II: Their Diversity in Space and Time.*Springer 2012, 427-440.

Amirlatifi, F., Soltani, N., Saadatmand, S., Shokravi, Sh. and Dezfulian, M. (2013) Crude Oil-induced Morphological and Physiological Responses in Cyanobacterium Microchaete tenera ISC13. Int. J. Environ. Res.,2013, 7(4):1007-1014

Antić, M.P., Jovancićević, B.S., Ilić, M., Vrvić, M.M. & Schwarzbauer, J. (2006) Petroleum pollutant degradation by surface water microorganisms. *Environ Sci Pollut Res Int.* 13, 320-327.

Ardelean, I.I. (2006) Biosensors with Cyanobacteria and Algae In Recent Advances on Applied Aspects of Indian Marine Algae with reference to Global Scenario. Vol II (Ed. A. Tewari) Central Salt & Marine Chemicals Research Institute, 87-103.

Ardelean, I.I. & Peschek, G.A. (2011) The Site of Respiratory Electron Transport in Cyanobacteria and Its Implication for the Photo-Inhibition of Respiration in Peschek, G. A., Obinger, C., Renger, G. (Eds.*), Bioenergetic processes of cyanobacteria – from evolutionary singularity to ecological diversity*, 131-136; Springer, New York; ISBN 978-9-400-703520.

Babaei, S., Najafi, F., Soltani, N., Khavari-Nejad, R. A. & Abbaspanah, B. (2013) Physiological Responses of *Anabaena* sp. ISC55 to Crude Oil and Its Potential for Biodegradation. *International Journal on Algae. 15*, 264-273.

Beaudoin, D.J., Carmichael, C.A., Nelson, R.K., Reddy, C.M., Teske, A.P. & Edgcomb, V.P. (2014) Impact of protists on a hydrocarbon-degrading bacterial community from deep-sea Gulf of Mexico sediments: A microcosm study. Deep Sea Research Part II: Topical Studies in Oceanography Available online 26 January 2014.

Bershova, O.I., Kopteva, Z.H. & Tansyurenko, E.V. (1968) The interplay between blue-green algae - the causative agent of water “bloom”- and bacteria. In Topachevsky AV (Ed) Tsedeveny vody, 159-171, Naukova Dumbka, Kiev, URSS.

Bryant, D.A. (ed.) (1994) The molecular biology of cyanobacteria, Kluwer Academic Publishers.

Cao, X., Xiong, Y., & Lund, J. (2013) The effects of micro-algae characteristics on the bioremediation rate of Deepwater Horizon crude oil, *Journal of Emerging Investigators.* 1-7.

Cerniglia, C.E., Van Baalen, C. & Gibson, D.T. (1980) Metabolism of naphthalene by the cyanobacterium Oscillatoria sp., strain JCM. *J Gen Microbiol. 116*, 485–594.

Cerniglia, C.E., Gibson, D.T. & Van Baalen, C. (1980) Oxidation of naphthalene by cyanobacteria and microalgae. *J Gen Microbiol. 116*, 495–500.

Chaillan, F., Gugger, M., Sailot, A., Coute, A. & Oudot, J. (2006) Role of cyanobacteria in the biodegradation of crude oil by a tropical cyanobacterial mat. *Chemosphere. 62*, 1574–1582.

Chavan, A. & Mukherji, S. (2008) Treatment of hydrocarbon-rich wastewater using oil degrading bacteria and phototrophic microorganisms in rotating biological contactor: effect of N: P ratio. *J Hazard Mater* 154, 63–72.

Chavan, A. & Mukherji, S. (2010) Effect of co-contaminant phenol on performance of a laboratory-scale RBC with algal-bacterial biofilm treating petroleum hydrocarbon-rich wastewater. *J Chem Technol Biotechnol.* 85, 851–859.

Chavan, A. & Mukherji, S. (2010) Response of an algal consortium to diesel under varying culture conditions. *Appl Biochem Biotechnol* 160, 719–729.

Chronopoulou, P.M., A. Fahy, F. Coulon, S. Paisse, M. Goni-Urriza, L. Peperzak, L. Acuna Alvarez, B.A. McKew, T. Lawson, K.N. Timmis, R. Duran, G.J.C. Underwood, T.J. McGenity (2012) Impact of a simulated oil spill on benthic phototrophs and nitrogenfixing bacteria in mudflat mesocosms. Environ. Microbiol. 2012. Doi:10.1111/j.1462-2920.2012.02864.x.

Coates, R.C., Podell, S., Korobeynikov, A., Lapidus, A. & Pevzner, P. (2014) Characterization of Cyanobacterial Hydrocarbon Composition and Distribution of Biosynthetic Pathways. PLoS ONE 9(1): e85140. doi:10.1371/journal.pone.0085140.

Cohen, Y. (2002) Bioremediation of oil by marine microbial mats. *Int Microbiol. 5*, 189–193.

Cole, J.K., Hutchison, J.R., Renslow, R.S., Kim, Y.M., Chrisler, W.B., Engelmann, H.E., Dohnalkova, A.C., Hu, D., Metz, T.O., Fredrickson, J.K. & Lindemann, S.R. (2014) Phototrophic biofilm assembly in microbial-mat-derived unicyanobacterial consortia: model systems for the study of autotroph-heterotroph interactions. *Front. Microbiol. 5*, 109. doi: 10.3389/fmicb.2014.00109.

Cravo-Laureau, C. & Duran, R. (2014) Marine coastal sediments microbial hydrocarbon degradation processes: contribution of experimental ecology in the omics'era. *Front. Microbiol. 5*, 39. doi: 10.3389/fmicb.2014.00039.

de Oteyza, T.G., Grimalt, J.O., Diestra, E., Sole, A. & Esteve, I. (2004) Changes in the composition of polar and apolar crude oil fractions under the action of Microcoleus consortia. *Appl Microbiol Biotechnol. 66*, 226–232.

Fathepure, B.Z. (2014) Recent studies in microbial degradation of petroleum hydrocarbons in hypersaline environments. *Frontiers in Microbiology*. 5, 1-16.

Gault, P.M. & Marler, H.J. (eds.) (2009) *Handbook on Cyanobacteria: Biochemistry, Biotechnology and Applications*. New York, NY: Nova Science Publishers.

Gaur, J. P. & Singh, A. K. (1990). Growth, Photosynthesis and Nitrogen Fixation of Anabaena doliolum Exposed to Assam Crude Extract. *Bulletin Environmental Contamination and Toxicology. 44*, 494-500.

Ghiţă, S., Sarchizian, I. & Ardelean, I.I. (2013) Microscopic investigation and automated image analysis of hydrocarbon- tolerant marine cyanobacteria mixed populations cultivated in the absence and presence of gasoline or diesel. *International journal of biology and biomedical engineering*. 7, 164-175.

Govindjee & Shevela, D. (2011) Adventures with cyanobacteria: a personal perspective. *Front. Plant Sci*. 2, 1-17.

Head, I.M. & Swannell, R.P. (1999) Bioremediation of petroleum hydrocarbon contaminants in marine habitats. *Curr Opin Biotechnol. 10*, 234–239.

Hernández-Prieto, M.A., Semeniuk, T.A. & Futschik, M.E. (2014) Toward a systems-level understanding of gene regulatory, protein interaction, and metabolic networks in cyanobacteria. *Front. Genet.* **5**, 191. doi: 10.3389/fgene.2014.00191

Ibraheem I.B.M, I. (2010) Biodegradability of Hydrocarbons by Cyanobacteria. *Journal of Phycology. 46*, 818-824.

Klug, J. & Markovetz, J. (1971) Utilization of aliphatic hydrocarbons by micro-organisms. *Adv Microb Physiol. 5*, 1–39.

Knowles, R. & Wishart, C. (1977) Nitrogen-fixation in Arctic marine sediments – effect of oil and hydrocarbon factions. *Environ Pollut. 13*, 133–149.

Kuritz T., Wolk C.P. (1995) Use of filamentous cyanobacteria for biodegradation of organic pollutants. *Appl Environ Microbiol* 61,234–238.

Krell, T., Lacal, J., Reyes-Darias, J. A., Jimenez-Sanchez, C., Sungthong, R. & Ortega-Calvo, J.J., (2013) Bioavailability of pollutants and chemotaxis. *Current Opinion in Biotechnology 24*, 451–456.

Lacal, J., Muñoz-Martínez, F., Reyes-Darias, J.A., Duque, E., Matilla, M. & Segura, A., (2011) Bacterial chemotaxis towards aromatic hydrocarbons in Pseudomonas. Environ Microbiol. 7, 1733-1744.

Lange, W. (1971) Enhancement of algal growth in Cyanophyta- bacteria systems by carbonaceous compounds. *Can. J. Microbiol. 17*, 303-314.

Lăzăroaie, M.M. (2010) Multiple responses of Gram-positive and Gram-negative bacteria to mixture of hydrocarbons. *Brazilian Journal of Microbiology, 41*(3), 649-667.

Leahy, J.G. & Colwell, R.R. (1990) Microbial degradation of hydrocarbons in the environment. *Microbiol Rev* 54(3), 305.

Llirós M., Gaju N., de Oteyza T. G., Grimalt J. O., Esteve I., Martinez-Alonzo M., (2008) Microcosm experiments of oil degradation by microbial mats. II. The changes in microbial species. *Sci. Total Environ*. 393, 39–49.

Marx, R.B. & Aitken, M. (2000) Bacterial chemotaxis enhances naphthalene degradation in a heterogeneous aqueous system. *Environ. Sci. Technol. 34*, 3379–3383.

McGenity, T.J., Folwell, B.D., McKew, B.A., and Sanni, G.O. (2012) Marine crude-oil biodegradation: a central role for interspecies interactions. *Aquat Biosys* 8,10.

Morales A.R., Paniagua-Michel J. (2013) Bioremediation of Hexadecane and Diesel Oil is Enhanced by Photosynthetically Produced Marine Biosurfactants. *J Bioremed Biodeg* S4: 005. doi:10.4172/2155-6199.S4-005.

Murínová, S. & Dercová, K. (2014) Response Mechanisms of Bacterial Degraders to Environmental Contaminants on the Level of Cell Walls and Cytoplasmic Membrane, *International Journal of Microbiology* Volume 2014, Article ID 873081, 16 pages.

Musat, F., Harder, J. & Widdel, F. (2006) Study of nitrogen fixation in microbial communities of oil-contaminated marine sediment microcosms. *Environ Microbiol. 8*, 1834–1843.

Paerl, H.W., Pinckney, J.L. & Steppe, T.F. (2000) Cyanobacterial–bacterial mat consortia:examining the functional unit of microbial survival and growth in extreme environments. *Environ.Microbiol. 2*, 11–26.

Peschek, G.A. (1996) Structure-function relationships in the dual-function photosyntheticrespiratory electron-transport assembly of cyanobacteria (blue-green algae). *Biochemical Society Transactions* 24, 729- 733.

Peschek, G.A., Obinger, C. & Renger, G. (eds.) (2011) Bioenergetic Processes of Cyanobacteria: From Evolutionary Singularity to Ecological Diversity. Dordrecht: Springer.

Radwan, S.S., Al-Hasan, R.H., Salamah, S. & Khanafer, M. (2005) Oil consuming microbial consortia floating in the Arabian Gulf. *Int Biodeter Biodeg 56*, 8–33.

Radwan, S.S. & Al-Hasan, R.H. Oil polution and cyanobacteria. In The Ecology of Cyanobacteria. Whiton B.A and Potts M Editors, 2000 Kluwer Academic Publishers, 307-319.

Radwan, S.S. & Al-Hasan, R.H. (2001a) Potential application of coastal biolfilm-coated gravel particles for treating oily waste. *Aqu Microb Ecol 23*, 113–117.

Radwan, S.S. & Al-Hasan, R.H. (2001b) Cyanobacteria and the Self-Cleaning of oily Environments. In: Kojima H, Lee YK (eds) *Photosynthetic microorganisms in Environmental Biotechnology. 12*, 181–193.

Radwan, S.S., Al-Hasan, R.H., Salamah, S. & Al-Dabbous, S. (2002) Bioremediation of oily sea water by bacteria immobilized in biofilms coating macroalgae. *Int Biodeter Biodeg. 50*, 55–59.

Radwan, S.S., Mahmoud, H., Khanafer, M., Al-Habib, A. & Al-Hasan, R. (2010) Identities of epilithic hydrocarbon-utilizing diazotrophic bacteria from the Arabian Gulf coasts, and their potential for oil bioremediation without nitrogen supplementation. *Microbiol Ecol. 60*, 354–363.

Raghukumar, C., Vipparty, V., David, J.J. & Chandramohan, D. (2001) Degradation of crude oil by marine cyanobacteria. *Appl Microbiol Biotechnol. 57*, 433–436.

Ramos, J. L., Duque, E., Gallegos, M.T., Godoy, P., Ramos-González, M. I., Rojas, A., Terán, W., & Segura, A. (2002) Mechanisms of solvent tolerance in Gram-negative bacteria. *Annual Review of Microbiology 56*, 743-768.

Raven J.A.,(2012) Carbon. In Whitton B.A., editor.*Ecology of cyanobacteria II:Their Diversity in Space and Time.*Springer 2012, 443-460.

Rhem, H.J. & Reiff, J. (1981) Mechanisms and occurrence of microbial oxidation of long-chain alkanes. *Adv Biochem Eng 19*, 175–215.

Safferman, R.S. & Morris, M.E. (1962) Evaluation of natural products for algicidal properties. *Appl. Microbiol. 10*, 289-292.

Safonova, E.T., I.A. Dmitrieva, K.V. Kvitko (1999) The interaction of algae with alcanotrophic bacteria in black oil decomposition. *Resour. Conserv. Recycl.* 27. 193–201.

Safonova, E., Kvitko, K.V., Iankevitch, M.I., Surgko, L.F., Afti, I.A., Reisser, W., (2004). Biotreatment of industrial wastewater by selected algal–bacterial consortia. *Eng. Life Sci. 4,* 347–353.

Sánchez, O., Diestra, E., Esteve, I. & Mas, J. (2005) Molecular characterization of an oil degrading cyanobacterial consortium. *Microb Ecol. 50*, 580–588.

Sauret, C., Christaki, U., Moutsaki, P., Hatzianestis, I., Gogou, A. (2012) Influence of pollution history on the response of coastal bacterial and nanoeukaryote communities to crude oil and biostimulation assays. *Marine Environmental Research* 79, 70-78.

Seckbach, J. (ed.) (2007) Algae and Cyanobacteria in Extreme Environments, Series: Cellular Origin, Life in Extreme Habitats and Astrobiology, Springer.

Sorkhoh, N., Al-Hasan, R., Radwan, S. & Höpner, T. (1992) Self-cleaning of the Gulf. *Nature 350*, 109.

Sorkhoh, N.A., Ali, N., Dashti, N., Al-Mailem, D.M., Eliyas, M. & Radwan, S.S. (2010a) Soil bacteria with the combined potential for oilutilization, nitrogen-fixation and mercury-resistance. *Int Biodeter Biodegr. 64*, 226–231.

Sorkhoh, N.A., Ali, N., Salamah, S., Eliyas, M., Khanafer, M. & Radwan, S.S. (2010b) Enrichment of rhizospheres of crop plants raised in oily sand with hydrocarbon utilizing bacteria capable of hydrocarbon consumption in nitrogen free media. *Int Biodeter Biodegr. 64*, 659–664.

Speight, (1999) The chemistry and technology of petroleum. Marcell Dekker, New York, 215-216.

Srivastava, A.K., Rai, A.N. & Neilan, B.A. (ed.) (2013) Stress Biology of Cyanobacteria: Molecular Mechanisms to Cellular Responses, CRC Press, Taylor and Francis Group.

Thomson, A.D. & Webb, K.L. (1984) The effect of chronic oil pollution on salt-marsh nitrogen fixation (acetylene reduction). *Estuaries. 7*, 2–11.

Toccalino, P.L., Johnson, R.L. & Boone, D.R. (1993) Nitrogen limitation and nitrogen fixation during alkane biodegradation in a sandy soil. *Appl Environ Microbiol. 59*, 2977–2983.

van Bleijswijk, J. & Muyzer, G. (2004) Genetic diversity of oxygenic phototrophs in microbial mats exposed to different levels of oil pollution. *Ophelia. 58*, 157–164.

Zhubanova, A.A., Ernazarova, A.K., Kaiyrmanova, G.K., Zayadan, B.K., Savitskaya, I.S., Abdieva, G.Zh., Kistaubaeva, A.S., & Akimbekov, N.Sh. (2013) Construction of Cyanobacterial–Bacterial Consortium on the Basis of Axenic Cyanobacterial Cultures and Heterotrophic Bacteria Cultures for Bioremediation of Oil Contaminated Soils and Water Ponds. *Russian Journal of Plant Physiology. 60,* 555–562.

Whitton, B.A. (ed.) (2012) *Ecology of Cyanobacteria II. Their diversity in Space and Time.* Springer Netherlands.

In: Cyanobacteria
Editor: Douglas Davison

ISBN: 978-1-61470-925-1

Chapter 3

MATHEMATICAL MODELS FOR PREDICTING THE CYANOTOXINS PRESENCE IN SEVERAL RESERVOIRS IN THE CANTABRIAN BASIN (NORTHERN SPAIN)

P. J. García Nieto[1], J. R. Alonso Fernández[2], C. Díaz Muñiz[2], F. Sánchez Lasheras[3], and F. J. de Cos Juez[4]

[1]Department of Mathematics, University of Oviedo, Oviedo, Spain
[2]Cantabrian Basin Authority, Ministry of Agriculture, Food and Environment, Oviedo, Spain
[3]Department of Construction and Manufacturing Engineering, University of Oviedo, Spain
[4]M ining Exploitation and Prospecting Department, University of Oviedo, Oviedo, Spain

ABSTRACT

Eutrophication is a global and serious problem in lakes and reservoirs all over the world. An obvious and risky symptom of eutrophication is the rapid growth and accumulation of phytoplankton (called blooms) leading to discoloration of the affected waters. Blooms are a prime agent of the water quality deterioration, deoxygenation of bottom waters (hypoxia and anoxia) and toxicity. Toxins produced by blooms can adversely affect animals and human health in waters used for recreational and drinking purposes. Numerous genera are able to form blooms. However, cyanobacteria are the most notorious blooms formers: they are adept at exploiting nutrient-enriched conditions. One of their major problems arises from their ability to produce toxins. Consequently, preventing eutrophication is an important issue in lakes and reservoir management. In order to develop an eutrophication (or cyanobacteria) prevention system, a good starting point can be the diagnosis of the situation in the area where the selected reservoirs in this study are located. The Trasona reservoir was used as a pilot reservoir in which a more exhaustive study was carried out: many chemical, physico–chemical and biological parameters were monitored for years in it, mathematical models were developed to forecast eutrophication (and cyanobacteria presence), and therefore the risk of cyanotoxins, etc. The aim of this work was to obtain predictive models able to perform an early detection of cyanotoxins by using as predictors different chemical, physico–chemical and biological measurements. The study highlighted the importance of the different kinds of cyanobacteria in cyanotoxins production by using the above-mentioned models. Moreover, the models obtained were tried in other fifteen Cantabrian Basin reservoirs to illustrate that they could be used to successfully predict the cyanobacteria presence in the reservoirs of the study area. Finally, conclusions of this innovative research are exposed.

Keywords: Statistical machine learning techniques, Eutrophication, Cyanobacteria, Cyanotoxins

1. INTRODUCTION

The quality of surface waters, especially lentic waters, is frequently affected by eutrophication: a process where water bodies receive excess nutrients, particularly phosphorus and nitrogen, that stimulate excessive plant growth. Consequently, ecosystem control mechanisms are not used [1-2].

An obvious and problematic symptom of eutrophication is the rapid growth and accumulation of phytoplankton (called blooms) leading to discoloration of the affected waters [1-3], loss of transparency, unpleasant tastes and odors, and toxicity by the presence of certain algae [1]. Although eutrophication is generally recognized as natural process in the aging of lakes, human activity speed up the process by increasing the rate at which nutrients enter the aquatic environment from surrounding water-shed [4] generating one of the most prevalent global problems of this era. Regarding the blooms, cyanobacteria are the most notorious blooms formers: they are adept at exploiting nutrient-enriched conditions [5]. Their ability to produce toxins called cyanotoxins, worsens the eutrophication problem since they are some of the most powerful natural poisons known, including potent neurotoxins, hepatotoxins, cytotoxins, and endotoxins [5]. Recreational exposure to cyanobacteria can result in gastro-intestinal and hayfever symptoms or pruritic skin rashes. There is some evidence that significant exposure to high levels of some species of cyanobacteria causes Lou Gehrig's disease [5]. Furthermore, wildlife could be also affected harmfully by this kind of polluted water. Most reported incidents of poisoning by microalgal toxins have occurred in freshwater environments, and they are becoming more common and widespread [2-5]. Consequently, cyanotoxins are a growing environmental and public health concern as well as a serious economic problem in local economies as a consequence of lost recreational opportunities of the visitors [6], the mortality in aquaculture organisms [7], etc. Efficient management and protection of water bodies against cyanobacterial is, therefore, of critical importance. Cyanobacteria have been extensively studied and numerous strategies have been tested to prevent or eliminate their blooms. However, an entirely satisfactory solution still remains unknown. Some of them, for example, decreasing the inputs of nutrients, are partial solutions that improve the situation but that do not solve the problem in a reasonable period of time. Other strategies have important disadvantages, such as algaecide applications because of the subsequent release of toxins [8], or the drinking water treatments to remove simultaneously all the cyanotoxins of the mixture but not affordable for some small treatment plants.

Therefore, efforts must be focused on the occurrence of cyanobacterial blooms. Despite its difficulty, it is very important to establish predictive models [9].

2. The Aim of the Present Research

In order to develop an eutrophication (or cyanobacteria) prevention system, a good starting point was the situation diagnosis in the area where the selected reservoirs for this study are located. Figure 1 shows the geographical limits of the Cantabrian Basin Authority where the selected reservoirs are located. Figure 2 shows the reservoirs studied. A pilot reservoir (see Figure 3) was subjected to a more exhaustive study in order to elaborate the eutrophication prevention and prediction system. The Trasona reservoir (located in Asturias, Northern Spain;

Figure 1. Geographical limits of Cantabrian Basin Authority: its boundary is limited by a black solid line.

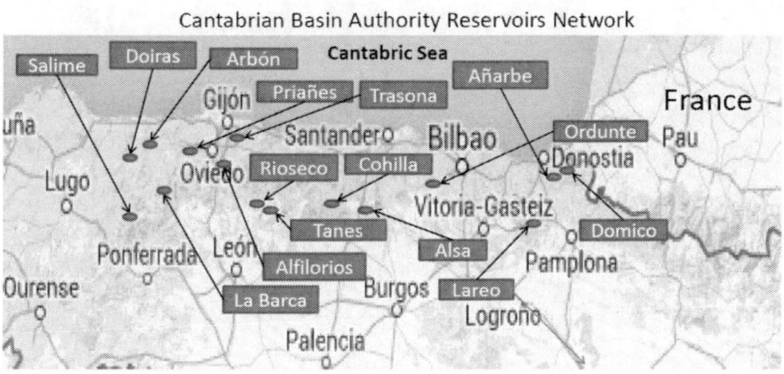

Figure 2. The reservoirs studied.

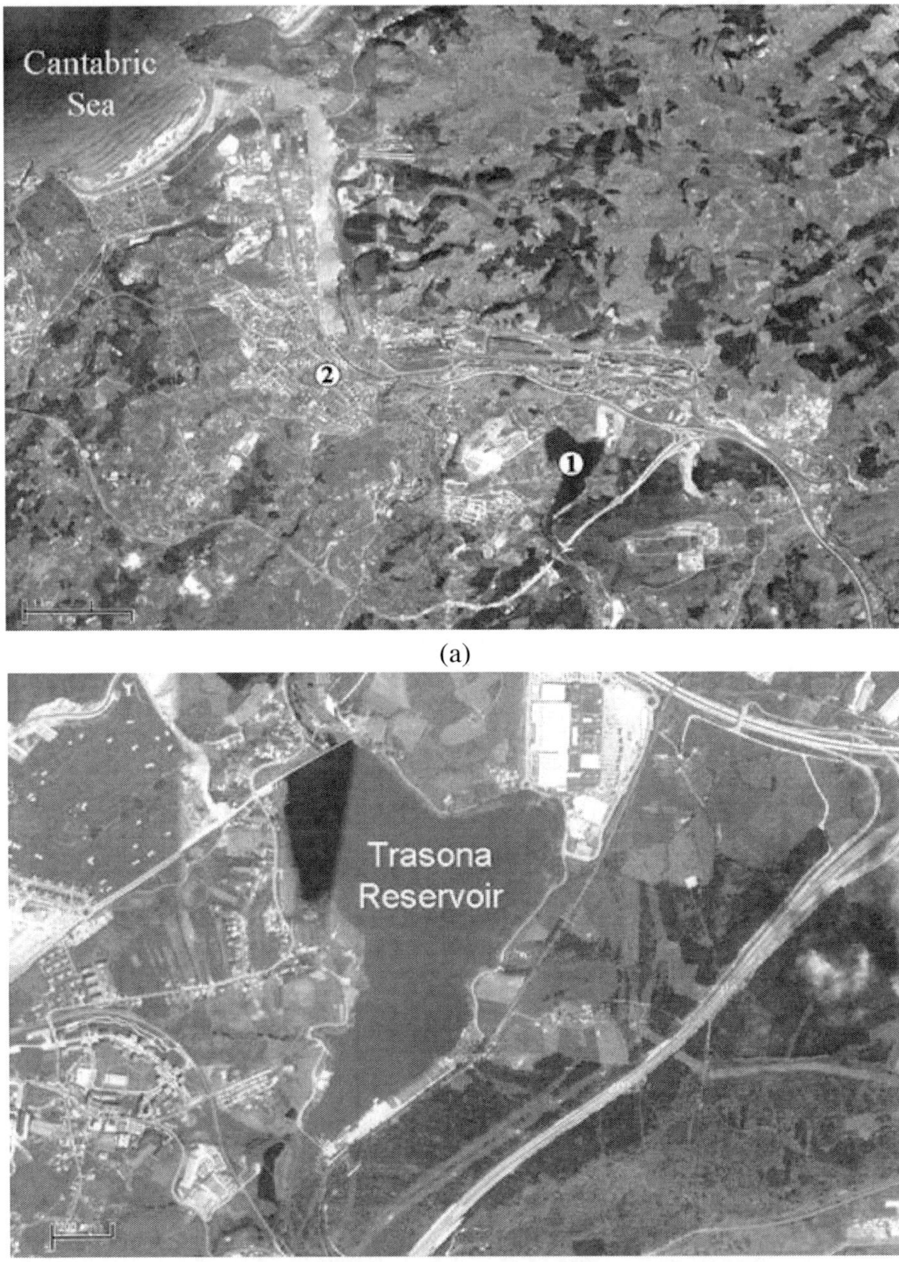

Figure 3. (a) Aerial photograph of the Avilés city (Northern Spain) (2) and Trasona reservoir (1); and (b) an aerial photograph of Trasona reservoir in great detail.

Figure 3(a)-(b)) was selected as a pilot reservoir and many chemical, physico-chemical and biological parameters were measured during years. Additionally, mathematical models were developed to forecast eutrophication (and cyanobacteria presence), and therefore the cyanotoxins risk. In our previous research work [10], a model based on the multivariate adaptive regression splines (MARS) technique was built to study cyanotoxins occurrence (dependent variable) from biological data set focused on the number of cells (the independent variables) in the Trasona reservoir. However, a more accurate analysis of this complex problem requires the consideration of the biovolume estimations and of some essential physico-chemical parameters (water temperature, turbidity, conductivity, pH, etc.) as independent variables. Microalgae have a wide range of shapes and size, from submicron species such as the picoplanktonic prochlorophytes to diatoms greater than 1 mm. in diameter [11]. Therefore, cell counts per se are inadequate as a measure of relative algal biomass. A standard biomass estimate is essential for comparing the relative contribution of different microalgae. As a result, depending on the equivalent geometric shapes for each microalgae set, different equations were used in order to estimate their biovolume [11].

Firstly, a more complete regression model based on the MARS technique [12-16] was performed to identify cyanotoxins in the Trasona reservoir (Asturias, Northern Spain) (see Figure 3). Secondly, a more complete regression model based on the combination of a genetic algorithm (GA) to reduce the initial total number of predicting variables was developed followed by the MARS technique's application [12-14] to identify the cyanotoxins presence in the above-mentioned reservoir as a function of the main predicting variables selected previously using a suitable genetic algorithm [16]. The study highlighted the importance of each kind of cyanobacteria in cyanotoxins production by using the above-mentioned models. Moreover, the models obtained were tried in other fifteen Cantabrian Basin reservoirs to illustrate their ability to be successfully used in prediction of the cyanobacteria occurrence in any reservoir within the study area.

3. Experimental Dataset

The studied reservoirs (artificial water bodies and heavily modified water bodies comparable to rivers) and artificial lakes (*Lareo*, *Domico* and *Alfilorios*) are shown in Table 1.

Table 1. Reservoirs studied

Name	Volume (hm^3)	Use	Geology
Alfilorios	9	Drinking Water Supply	Calcareous
Alsa	23	Production of electricity	Calcareous
Añarbe	44	Drinking Water Supply	Siliceous
Arbón	33	Production of electricity	Siliceous
Cohilla	12	Production of electricity	Calcareous
Domico	6	Drinking Water Supply	Siliceous
Doiras	15	Production of electricity	Siliceous
La Barca	32	Production of electricity	Siliceous
Lareo	3	Drinking Water Supply	Siliceous
Ordunte	22	Drinking Water Supply	Calcareous
Priañes	2	Production of electricity	Siliceous
Rioseco	4	Drinking Water Supply and Production of electricity	Calcaleous
Salime	266	Production of electricity	Siliceous
Tanes	33	Drinking Water Supply and Production of Eelectricity	Calcareous
Trasona	4	Industrial Supply and Sport Utilization	Calcareous

These fifteen reservoirs were sampled twice in summer (June and September) from 2006 to 2012. As above-mentioned, in this study, the Trasona reservoir was taken as the pilot reservoir. The data set used for both the regression model based on the MARS technique and the hybrid GA-MARS model, were collected over six years (2006 to 2011) from lots of samples in the Trasona reservoir. The total number of data processed was about 151 values. The time distribution of this data set depended on the cyanobacterial blooms occurrence. In this way, four samples per month were collected in the months with the cyanobacterial blooms occurrence (July, August, September and October of 2006, 2007 and 2008). However, in November of 2006, 2007 and 2008, the sampling was carried out three times per month. For the remaining months, the reservoir was sampling twice per month, except in 2011 where only two samples were collected: in July and September. The samples were taken following the sampling protocols for lakes and reservoirs of the Spanish Ministry of Agriculture, Food and Environment, which are consistent with the guidelines established by the European Union and international agencies dealing with these issues [2-5]. A Niskin hydrographic bottle was used to collect the water samples. The samples were taken at different depths in the euphotic zone [2-5], that is to say, the layer closer to the surface that receives enough light so that photosynthesis to take place. Its depth, determined by the transparency of the water, is measured using a Secchi disk. In this sense, the euphotic zone depth is equal to 2.5 times the depth at which the pattern on that disk is no longer visible when it is lowered down in the water. Biological parameters were expressed in biovolume (cubic millimeters per liter) of phytoplankton species.

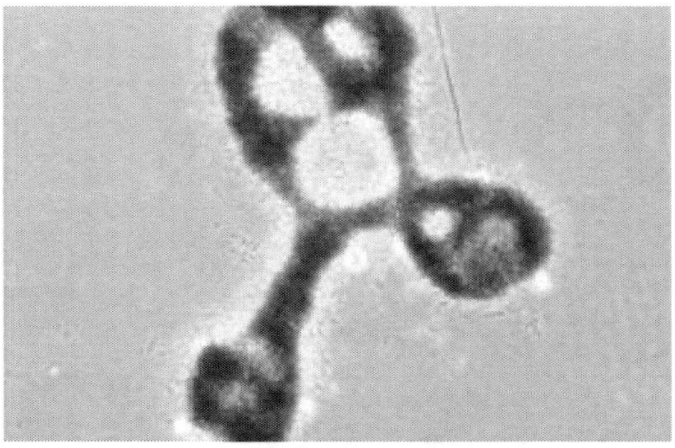

(a)

(b)

Figure 4. The most abundant cyanobacteria species in the studied reservoirs: Microcystis aeruginosa (a) and Woronichinia naegeliana (b).

The values of phytoplankton, cyanotoxins and chlorophyll concentrations as well as other physicochemical parameters, were determined from a sample composed of five homogeneous subsamples obtained with the hydrographic bottle at various equidistant depths in the euphotic zone [2-5].

In this work, the two dominant species in the cyanobacteria community were taken into account: Microcystis aeruginosa (Figure 4(a)) and Woronichinia naegeliana. (Figure 4 (b)).

To obtain the dependence relationship of cyanotoxins (output variable, expressed in micrograms per liter), the following eight biological and fifteen physico-chemical variables were considered:

➢ Biological input parameters:

- *Microcystis aeruginosa* (mm^3/L): a type of harmful blue-green algae which is also referred to as colonial cyanobacteria (see Figure 4(a)).
- *Woronichinia naegeliana* (mm^3/L): a kind of cyanobacteria in low trophic status waters (Figure 4(b)).
- Other cyanobacteria (mm^3/L): all cyanobacteria excluding the two previous ones. Examples of these species may include some potentially toxic species such as *Microcystis flos-aquae*, *Microcystis novacekii*, *Anabaena flos-aquae* and *Anabaena crassa*.
- Diatoms (mm^3/L): a major group of algae and one of the most common phytoplankton types.
- Chrysophytes (mm^3/L): small yellowish brown flagellates that can be found singly or in a colony.
- Chlorophytes (mm^3/L): a highly paraphyletic group of all green algae within the green plants group.
- Other phytoplankton species (mm^3/L): all phytoplankton species excluding the previous ones.
- Chlorophyll concentration (μg/L): a measure of the density of an extremely important biomolecule (chlorophyll), critical in photosynthesis since it allows plants to obtain energy from light.

➢ Physico-chemical input parameters:

- Water temperature (ºC): the temperature of water reservoir. It affects the solubility of many chemical compounds and therefore can influence the pollutants effects on aquatic life.
- Ambient temperature (ºC): the temperature of reservoir surroundings (air temperature).
- Secchi disk depth (m): the depth at which the pattern on the Secchi disk (a circular disk with alternating black and white quadrants, mounted on a pole or line) is no longer visible from the surface when it is lowered down in the water. It is a measurement of the transparency of the water. The transparency decreases as the algal abundance increases and as a consequence, the turbidity increases.

- Turbidity (NTU): the water cloudiness or haziness caused by individual particles (suspended solids). The turbidity measurements is a key test for water quality control.
- Total phosphorus (mg P/L): the amount of both inorganic and organic forms of phosphorus. Phosphorus can be present as dissolved or particulate matter.
- Phosphates concentration (mg PO_4^{3-}/L): the amount of an often limiting nutrient in ecological environments. Its availability may govern the growth rate of the aquatic organisms.
- Total nitrogen concentration (mg N/L): the amount of the organically and inorganically bounded nitrogen. Organic nitrogen includes all organic compounds such as proteins, polypeptides, amino acids, and urea, all of them essential to Earth´s life.
- Nitrate concentration (mg NO_3^-/L): the amount of a form of nitrogen resulting from the complete oxidation of nitrogen compounds. Its excess may result in phytoplankton or macrophyte proliferations.
- Nitrite concentration (mg NO^{2-}/L): this form of nitrogen can also be used as a source of nutrients for plants. Nitrite is toxic to aquatic life at relatively low concentrations.
- Ammonium/ammonia concentration (mg/L): the amount of the most reduced inorganic nitrogen form in water. Ammonia excess contributes to eutrophication of water bodies. It can be easily oxidized to nitrate in oxidizing environments.
- Dissolved oxygen concentration (mg O_2/L): the amount of dissolved oxygen in water. It is essential to the respiratory metabolism of most aquatic organisms. It affects the nutrients' solubility and availability, and therefore the aquatic ecosystems productivity.
- Conductivity (μS/cm): the electrolyte solution ability to conduct electricity. It is related to the salt content in water.
- Alkalinity (mg $CaCO_3$/L): the solution ability to neutralize acids to the equivalence point of carbonate.
- Calcium concentration (mg/L): amount of an essential metal for living organisms, in particular in cell physiology.
- pH: the acidity or basicity of the aqueous solution. Lethal pH effects on aquatic life occur below pH 4.5 and above pH 9.5.

Since all the water surface reservoirs had less than 50 hectares, one unique sampling was taken at the deepest point. Simultaneously, the multiparameter probe was used to obtain the temperature, conductivity, turbidity, pH, redox potential and dissolved oxygen profiles, in order to identify the pattern of stratification. This dataset is included in the subsequent mathematical treatment, along with all of the data from the analytical result of the samplings.

"Chlorophyll *a*" was quantified by spectrophotometry, after the pigments extraction. Phytoplankton composition was determined by Utermolhl's method with inverted microscope.

The procedure includes methodological guidelines to:

- identify the necessary laboratory equipment.
- prepare the samples for their examination using the inverted light microscope.
- identify the taxa and count their cells.

- apply statistical methods to optimize the counting method.
- record both the data and the samples.

This calculation is based on the biovolumes of the different algal groups considered for each of the indexes. The cyanobacteria percentages were calculated from the corresponding biovolume.

The cells inventories were obtained by an inverted microscope on settled samples. The cyanotoxins were quantified by high-performance liquid chromatography (HPLC) technique [2-5].

4. Mathematical Model

4.1. Multivariate Adaptive Regression Spline

The MARS model of a dependent variable with M basis functions (terms) can be written as [10, 13-16]:

$$\hat{y} = \hat{f}_M(\vec{x}) = c_0 + \sum_{m=1}^{M} c_m B_m(\vec{x}) \quad (1)$$

where $\hat{y}$ is the dependent variable predicted by the MARS model, c_0 is a constant, $B_m(\vec{x})$ is the m-th basis function, which may be a single spline basis functions, and is the coefficient of the m-th basis functions. In order to determine which basis functions should be included in the model, MARS utilizes the generalized cross-validation (GCV) [6,8-12]. In this way, the GCV is the mean squared residual error divided by a penalty dependent on the model complexity. The GCV criterion is defined in the following way [13-16]:

$$GCV(M) = \frac{\frac{1}{n}\sum_{i=1}^{n}\left(y_i - \hat{f}_M(\vec{x}_i)\right)^2}{\left(1 - C(M)/n\right)^2} \quad (2)$$

where $C(M)$ is a complexity penalty that increases with the number of basis functions in the model and which is defined as [13-16]:

$$C(M) = (M+1) + dM \quad (3)$$

where M is the number of basis functions in Eq. (4), and the parameter d is a penalty for each basis function included into the model. It can be also regarded as a smoothing parameter. Large values of d lead to fewer basis functions and therefore smoother function estimates. For more details about the selection of the d parameter, see the references [13-16]. In our studies, the parameter d equals 2, and the maximum interaction level of the spline basis functions is restricted to 3.

4.2. Genetic Algorithms

The genetic algorithms (GAs) are based upon Darwin's Theory of Evolution. The genetic algorithms are modelled on a relatively simple interpretation of the evolutionary process [12, 17-19]. However, it has proven to be a reliable and powerful optimization technique in a wide variety of applications. Holland [12] in 1975 was the first to propose the use of genetic algorithms for problem-solving. The GA uses the current population of strings to create a new population whereby the strings in the new generation are on average better than those in the current population; the selection depends on their fitness value.

The selection process determines which string in the current will be used to create the next generation. The crossover process determines the actual form of the string in the next generation. Weak individuals are discarded and only the strongest survive. In this way, how do they work?

- Initialization: Initially many individual solutions are randomly generated to form an initial population. The population size depends on the nature of the problem, but typically contains hundreds or even thousands of possible solutions. Traditionally, the population is generated randomly, covering the entire range of possible solutions (the search space) [17-19]. Occasionally, the solutions may be "seeded" in areas where optimal solutions are likely to be found.
- Evaluation: An evaluation function is applied in order to know the goodness of each of the solutions of the population [17-19].
- Stop criterion: The GA will stop when the optimum solution is found or after a certain number of iterations/generations. If the stop criterion is not accomplished then a new iterative loop is carried out [17-19].
- Selection: During each successive generation, a proportion of the existing population is selected to breed a new generation. Individual solutions are selected through a fitness-based process, where fitter solutions (as measured by a fitness function) are typically more likely to be selected. Certain selection methods rate the fitness of each solution and preferentially select the best solutions. Other methods rate only a random sample of the population, as this process may be very time-consuming. The fitness function, f, maps a chromosome representation into a scalar value so that represents the data type of the elements of an dimensional chromosome [17-19]:

$$f : \Gamma^{n_x} \rightarrow \Re \tag{4}$$

- Crossover: In genetic algorithms, crossover is a genetic operator used to vary the programming of a chromosome or chromosomes from one generation to the next. It is analogous to reproduction and biological crossover, upon which genetic algorithms are based. Crossover operators can be divided into three main categories based on the arity (i.e. the number of parents used) of the operator. This gives rise to three main classes of crossover operators [17-19]: (1) asexual, where an offspring is generated from one parent; (2) sexual, where two parents are used to produce one or two offspring - the operator employed in the present research - and (3) multi-recombination, where more than two parents are used to produce one or more offspring.

- Mutation: A genetic operator, used to maintain genetic diversity from one generation of a population of algorithm chromosomes to the next. It is analogous to biological mutation. Mutation is used in support of crossover to ensure that the full range of allele is accessible for each gene. Mutation is applied at a certain probability, p_m, to each gene of the offspring, $\tilde{\mathbf{x}}_i(t)$, to produce the mutated offspring $\mathbf{x}_i(t)$. The mutation probability, also referred to as the mutation rate, is usually a small value, $p_m \in [0,1]$, to ensure that good solutions are not distorted too much. Given that each gene is mutated at probability p_m, the probability that an individual will be mutated, taking into account that the individual contains n_x genes, is given by [17-19]:

$$Prob(\tilde{\mathrm{x}}_i(t) \text{ is mutated}) = 1-(1-p_m)^{n_x} \tag{5}$$

- Replacement: the least-fit population is replaced with new individuals [13-15].

5. Results

The biological and physico-chemical input variables considered in this work are shown in Tables 2 and 3, respectively. One of the variables is the product of two variables: M. aeruginosa multiplied by W. naegeliana. It represents the coexistence of these two cyanobacteria species in order to reproduce their dynamics without intervention of external factors.

This mathematical formulation adds a multiplicative additional term to take into account the interaction between both species according to a more realistic modelling in Biology [20-21].

Table 2. Set of biological input variables used in this study

Biological input variables	Name of the variable
Microcystis aeruginosa (mm^3/l)	Microcystis_aeruginosa
Woronichinia naegeliana (mm^3/l)	Woronichinia_naegeliana
Other cyanobacteria (mm^3/l)	Other_species_Cyanobacteria
Diatoms (mm^3/l)	Diatoms
Chrysophytes (mm^3/l)	Chrysophytes
Chlorophytes (mm^3/l)	Chlorophytes
Other species of the phytoplankton (mm^3/l)	Other_phyto
Microcystis aeruginosa × Woronichinia naegeliana (synergistic interaction variable) (mm^6/l^2)	Microcys_×_Worochinia
Chlorophyll concentration (μ g/l)	Chlorophyll

Table 3. Set of physico-chemical input variables used in this study

Physico-chemical input variables	Name of the variable
Water temperature (ºC)	Water_temperature
Ambient temperature (ºC)	Ambient_temperature
Secchi disk depth (m)	Secchi_disk_depth
Turbidity (NTU)	Turbidity
Total phosphorus (mg P/l)	Total_phosphorus
Phosphates concentration (mg PO_4^{3-}/l)	Phosphates_concentration
Total nitrogen concentration (mg N/l)	Total_nitrogen_concentration
Nitrate concentration (mg NO^{3-}/l)	Nitrate_concentration
Nitrite concentration (mg NO^{2-}/l)	Nitrite_concentration
Ammonium concentration (mg/l)	Ammonium_concentration
Dissolved oxygen concentration (mg O_2/l)	Dissolved_oxygen_concentration
Conductivity (μ S/cm)	Conductivity
Alkalinity (mg $CaCO_3$/l)	Alkalinity
Calcium concentration (mg/l)	Calcium_concentration
pH values	pH_ values

In order to select the model that best fits the experimental data, the coefficient of determination R^2 [13-16] was considered. This ratio indicates the proportion of total variation in the dependent variable explained by the model (cyanotoxins in this case).

A coefficient of determination value of 1.0 indicates that the regression curve fits the data perfectly.

In this work, for the fitted hybrid GA-MARS model the coefficient of determination and correlation coefficient were 0.96 and 0.98, respectively, indicating a better agreement with the observed data using the hybrid improved model [10] than using the MARS model-2 [15] whose coefficient of determination and correlation coefficient were 0.94 and 0.97, respectively (Table 4) or the fitted MARS model-1 [10] with a coefficient of determination and correlation coefficient of 0.84 and 0.92, respectively (see Table 4).

In attempting to model real-world problems or concepts using computational methods, the selection of an appropriate representation is of considerable importance [13-14]. The selection of features can have a considerable impact on the effectiveness of the overall resulting regression algorithm: the hybrid GA-MARS model. The main purpose of this selection is to reduce the number of features used in regression maintaining an acceptable accuracy.

The GA used to build the model is able to select 6 main predicting variables from the 24 initial predicting variables.

Table 4. Coefficients of determination and correlation coefficients for the three models obtained: MARS model-1, MARS model-2, and hybrid GA-MARS

Model	Coefficients of determination /correlation coefficients
MARS model-1 (2011)	0.84/0.92
MARS model-2 (2012)	0.94/0.97
Hybrid GA-MARS model (2013)	0.96/0.98

Table 5. Variables significance in MARS model-2 (2012) according to criteria Nsubsets, GVC and RSS

Variables	Nsubsets	GCV	RSS
Microcys_x_Woronichinia	11	100	100
pH values	11	100	100
Alkalinity	10	36.2	37.8
Woronichinia_naegeliana	9	29.4	31.0
Turbidity	8	25.1	26.5
Water temperature	7	21.2	22.6

Table 6. Variable significance in the model according to criteria Nsubsets, *GCV* and *RSS* (GA-MARS)

Variable	Nsubsets	*GCV*	*RSS*
Microcys_×_Worochinia	33	100	100
Turbidity	32	70.5	69.5
Total_phosphorus	30	54.5	52.0
Alkalinity	25	26.8	23.8
Woronichinia_naegeliana	21	20.3	16.9
Water_temperature	18	15.4	12.4

The variable significance in the MARS model-2 is shown in Table 5. The most significant variable in cyanotoxins (output variable) prediction was the product of the M. aeruginosa concentration by the W. naegeliana concentration (Microcys_x_Woronichinia), followed by pH, alkalinity, W. naegeliana, turbidity and water temperature [15]. The mentioned product is a measure of the interaction between input variables M. aeruginosa and W. naegeliana. This interaction is known as synergy or synergistic behavior and it was also the most significant variable when only biological parameters were considered in the MARS model (MARS model-1) [10]

Therefore, as it is shown in Table 5, the twenty four original variables of this nonlinear complex problem were reduced to six main variables with minimal loss of information.

In summary, once the six main variables were selected by using an appropriate GA, a regression model based on multivariate adaptive regression splines (MARS technique) was developed with success in the Trasona reservoir in order to determine its cyanotoxins concentration.

To apply the preventing cyanobacteria proliferation and cyanotoxins occurrence system in any of the 15 studied reservoirs, summarized in Table 7 (with a correlation coefficient of 0.97), and in Table 8 (with a correlation coefficient of 0.98), the mean values of the surveys conducted from 2006 to 2011 (see Table 7) were used. The biological and physico-chemical parameters were:

- Microcystis_x_Woronichinia biovolume (in mm^3/L).
- Turbidity (NTU).
- pH (pH values).

- Total phosphorus (mg P/ L).
- Alkalinity (mg $CaCO_3$/L).
- Woronichinia naegeliana: biovolume (in mm^3/L).
- Water temperature (°C).

These fifteen reservoirs were sampled twice in the summer (June and September) for seven years (2006 to 2012). The mean values of the seven variables in 2012 for the studied reservoirs are shown in Table 8.

Table 7. Mean values of the seven variables in the period 2006-2011 of the reservoirs studied

Reservoirs/Variables	Mic._x_ Woron.	Tur b.	pH	T. Phosp.	Alkal.	W. naegel.	W. Temp
Alfilorios	0.0000	0.6	8.2	0.010	2.77	0.0000	17.3
Alsa	0.0034	3.9	8.1	0.035	1.05	0.0290	16.2
Añarbe	0.0000	1.5	7.8	0.010	0.38	0.0000	16.6
Arbón	0.0019	3.6	8.1	0.011	0.35	0.0100	17.5
Cohilla	0.0011	7.2	7.6	0.012	0.73	0.0072	16.8
Doiras	0.0009	4.6	8.2	0.040	0.36	0.0250	17.8
Domico	0.0000	2.1	7.4	0.010	0.63	0.0000	16.1
La Barca	1.0419	5.1	8.4	0.035	1.48	0.7311	20.2
Lareo	0.0000	1.9	8.3	0.010	1.51	0.0000	17.7
Ordunte	0.0000	1.4	8.1	0.010	1.44	0.0000	16.6
Priañes	0.0008	4.4	8.3	0.025	1.87	0.0089	17.4
Rioseco	0.0000	1.5	7.7	0.010	1.77	0.0000	16.4
Salime	0.0021	3.7	7.9	0.021	0.45	0.0054	18.2
Tanes	0.0000	2.2	7.8	0.010	1.80	0.0000	17.1
Trasona	13.3513	7.6	8.6	0.068	1.75	2.2482	19.1

Table 8. Mean values of the seven variables in the year 2012 of the reservoirs studied

Reservoirs/Variables	Mic._x_ Woron.	Tur b.	pH	T. Phosp.	Alkal.	W. naegel.	W. Temp
Alfilorios	0.0000	1.4	8.1	0.010	1.47	0.0000	16.6
Alsa	0.0005	5.6	8.2	0.019	1.06	0.0010	17.3
Añarbe	0.0000	3.2	7.6	0.000	0.36	0.0000	16.7
Arbón	0.0011	3.9	7.5	0.012	0.32	0.0041	17.4
Cohilla	0.0003	4.3	7.5	0.013	0.81	0.0019	16.3
Doiras	0.0020	5.1	8.1	0.011	0.32	0.0035	19.1
Domico	0.0000	2.9	7.3	0.010	0.74	0.0000	16.3
La Barca	0.0114	6.1	8.2	0.030	0.79	0.0705	19.1
Lareo	0.0000	2.9	7.7	0.010	1.14	0.0000	17.8
Ordunte	0.0000	0.6	7.6	0.010	1.44	0.0000	17.5
Priañes	0.0011	4.1	8.0	0.022	1.52	0.0006	18.0
Rioseco	0.0000	1.2	7.5	0.010	1.78	0.0000	18.4
Salime	0.0008	3.7	8.1	0.015	0.39	0.0010	19.2
Tanes	0.0000	0.6	7.7	0.010	1.79	0.0000	19.0
Trasona	0.0705	8.2	8.3	0.071	1.57	0.1507	19.9

Table 9. Reservoir classification as a function of the toxicity risk caused by cyanobacterial bloom

Type of risk	Reservoirs included in this risk
High risk	Trasona La Barca
Moderate risk	Priañes Arbón Doiras Salime Alsa Cohilla
Low risk	Alfilorios Añarbe Domico Lareo Ordunte Rioseco

The cyanobacterial bloom toxicity risk in the 15 studied reservoirs was based on the following three risk levels: high, moderate and low risk (see Table 9).

Trasona and La Barca reservoirs are the two ones with a high risk. In fact, the Trasona reservoir is a eutrophic ecosystem characterized by cyanobacteria outcrops in certain periods of time and also by a variable cyanotoxins production, mainly mycrocistins, in some occassions [10]. At the same time, the values of the turbidity and total phosphorus indicate a high risk of toxicity by cyanobacterial blooms. The La Barca reservoir is a meso-eutrophic ecosystem also characterized by cyanobacteria outcrops but never by the cyanotoxins production. The values of the turbidity and total phosphorus indicated a high toxicity risk by cyanobacterial blooms.

Priañes, Arbón, Doiras, Salime, Alsa and Cohilla reservoirs were characterized by a moderate cyanobacteria occurrence. Turbidity and total phosphorus values indicate also a moderate toxicity risk by cyanobacterial blooms (see Table 9).

Finally, Alfilorios, Añarbe, Domico, Lareo, Ordunte, Tanes and Rioseco reservoirs were characterized by a high cyanobacteria occurrence. Turbidity and total phosphorus values indicate a high toxicity risk by cyanobacterial blooms (see Table 9).

DISCUSSION AND CONCLUSION

The MARS model [10-15] and MARS model in combination with the genetic algorithms (GAs), are suitable tools in modelling and assessment of singular problems, such as cyanotoxin diagnosis in reservoirs. The Trasona reservoir was used as a pilot reservoir: many chemical, physico-chemical and biological parameters were measured for years in it and mathematical models were developed to forecast its eutrophication/cyanobacteria occurrence and therefore cyanotoxins risk. Biological input variables (phytoplankton species expressed in biovolume and the chlorophyll concentration) in combination with the most important physico-chemical parameters were used to fully characterize the problem.

A correlation coefficient of 0.97 was obtained when the MARS technique was applied to the experimental data set [15]. This value was increased to 0.98 applying the hybrid GA–MARS technique to the experimental data set [16]. The results are consistent with previous studies [12-21].

One of the main findings in this study, was to set the significance order of the variables involved in the prediction of the cyanotoxins occurrence. Specifically, the synergetic effect of the M. aeruginosa and W. naegeliana obtained by their concentrations product, is the most influential variable in cyanotoxins production. The second variable is the turbidity followed by pH, total phosphorus, alkalinity, W. naegeliana and water temperature. Finally, the methodology developed in this research work was applied to other fifteen reservoirs estimating their cyanobacterial bloom toxicity risks based on the following three risk levels: high, moderate and low risk (Table 9). The results were in a good agreement with the historical observations.

ACKNOWLEDGMENTS

The authors wish to acknowledge the computational support provided by the Departments of Mathematics, Construction and Mining Exploitation at University of Oviedo as well as pollutant data in the reservoirs studied supplied by the Cantabrian Basin Authority (Ministry of Agriculture, Food and Environment of Spain). Finally, we would like to thank Anthony Ashworth for his revision of English grammar and spelling of the manuscript.

REFERENCES

[1] Reynolds, C.S. Ecologogy of Phytoplnkton. New York: Cambridge University Press; 2006.

[2] Quesada, A., Moreno, E., Carrasco, D., Paniagua, T., Wormer, L., de Hoyos, C., and Sukenik, A. (2006). Toxicity of Aphanizomenon ovalisporum (cyanobacteria) in a spanish water reservoir. *European Journal of Phycology*, 41(1), 39-45.

[3] Vasconcelos, V. (2006). Eutrophication, toxic cyanobacteria and cyanotoxins: when ecosystems cry for help. *Limnetica*, 25(1-2), 425-432.

[4] Dasí, M.J., Miracle, M.R., Camacho, A., Soria, J.M. and Vicente, E. (1998). Summer phytoplankton assemblages across trophic gradients in hard-water reservoirs. *Hydrobiologia*, 369-370, 27-43.

[5] Fogg, G.E., Stewart W.D.P, Fay P., and Walsby, A.E. The Blue-green Algae. London: Academic Press; 1973.

[6] Steffensen, D.A. Economic cost of cyanobacterial blooms, In: Hudnell, K.H. (Ed.), Cyanobacterial harmful algal blooms: state of the science and research needs, Springer, New York, 2008, pp. 855–865.

[7] Smith, J.L., Boyer, G.L., Zimba, P.V. (2008). A review of cyanobacterial odorous and bioactive metabolites: Impacts and management alternatives in aquaculture. *Aquaculture*, 280(1–4), 5–20.

[8] Jones, G.J., Orr, P.T. (1994). Release and degradation of microcystin following algicide treatment of a Microcystis aeruginosa bloom in a recreational lake, as determined by HPLC and protein phosphatase inhibition assay. *Water Resources,* 28, 871–876.

[9] Merel, S., Walker, D., Chicana, R., Snyder, S., Baurès, E., Thomas, O. (2013). State of knowledge and concerns on cyanobacterial blooms and cyanotoxins. *Environment International,* 59, 303–327

[10] García Nieto, P.J., Sánchez Lasheras, F., de Cos Juez, F.J., and A. lonso Fernández, J.R. (2011). Study of cyanotoxins presence from experimental cyanobacteria concentrations using a new data mining methodology base don multivariate adaptive regression splines in Trasona reservoir (Northern Spain). *Journal of Hazardous Materials*, 195, 414-421.

[11] Hillebrand, H., Dürsrlen, C.D., Kirschtel, D., Pollingher, U., and Zohary, T. (1999). Biovolume calculation for pelagic and benthic microalgae. *Journal of Phycology*, 35, 403-424.

[12] Engelbrecht, A.P. Computational Intelligence: An Introduction. New York: Wiley; 2007.

[13] Friedman, J.H. (1991). Multivariate adaptive regression splines. *Annals of Statistics*, 19(1), 1-141.

[14] Hastie, T., Tibshirani, R. and Friedman, J. The Elements of Statistical Learning: Data Mining, Inference, and Prediction. New York: Springer; 2009.

[15] García Nieto, P.J., Alonso Fernández, J.R., Sánchez Lasheras, F., de Cos Juez, F.J., and Díaz Muñiz, C. (2012). A new improved study of cyanotoxins presence from experimental cyanobacteria concentrations in the Trasona reservoir (Northern Spain) using the MARS technique. *Science of the Total Environment*, 430, 88-92.

[16] Alonso Fernández, J.R., Díaz Muñiz, C., García Nieto, P.J., de Cos Juez, F.J., Sánchez Lasheras, F., and Roqueñí, M.N. (2013). Forecasting the cyanotoxins presence in fresh waters: A new model base don genetic algorithms combined with the MARS technique. *Ecological Engineering*, 53, 68-78.

[17] Holland, J.H. A*daptation in Natural and Artificial Systems. Boston: Ann Arbor,* The University of Michigan Press; 1975.

[18] Goldberg, D.E. *Genetic Algorithms in Search, Optimization and Machine Learning.* New York: Addison-Wesley; 1989.

[19] Davis, L. *Handbook of Genetic Algorithms.* New York: Van Nostrand Reinhold; 1991.

[20] Allman, E.S., and Rhodes, J.A. Mathematical Models in Biology: An Introduction. New York: Cambridge University Press; 2003.

[21] Barnes, D.J., and Chu, D. *Introduction to Modeling for Biosciences.* New York: Springer; 2010.

Reviewed by:

Prof. Dr. Pedro Arias,
Departament of Natural Resources and Environment,
University of Vigo, 36310 Vigo, Spain. (Email: parias@uvigo.es).

In: Cyanobacteria
Editor: Douglas Davison

ISBN: 978-1-61470-925-1

Chapter 4

PCR AND REAL-TIME PCR FOR IDENTIFICATION AND QUANTIFICATION OF CYANOBACTERIA TOXIC SPECIES AND CYANOTOXIN-PRODUCING GENOTYPES IN MACAU MAIN STORAGE RESERVOIR

Inchio Lou[1,*], Shuwen Xu[1], Weiying Zhang[1], Yijun Kong[2] and Kai Meng Mok[1]

[1]Department of Civil and Environmental Engineering, Faculty of Science and Technology, University of Macau, Macau SAR, China

[2]Laboratory and Research Center, Macao Water Co. Ltd., Macau SAR, China

ABSTRACT

Freshwater algal blooms have become a growing concern in Macau Storage Reservoir (MSR), which were caused by a high level of cyanobacteria, particularly Microcystis spp. and C. raciborskii that can produce microcystin and cylindrospermopsin, respectively. Long-time exposure to these cyanotoxins may affect public health, thus reliable detection and quantification of the algae species were challenging in water quality management. The aim of this study was to develop accurate and sensitive molecular methods on detecting cyanobacterial species and cyanotoxins-producing genotypes. The cyanobacteria, Microcystis spp. and C. raciborskii were identified and quantified by polymerase chain reaction (PCR), multiplex PCR and real-time quantitative PCR (qPCR) techniques. As well, the cyanotoxins-producing genotypes were also analyzed by PCR and qPCR using cylindrospermopsin polyketide synthetase gene (pks) and microcystin synthetase genes (mcys). The related water eutrophication, phytoplankton community diversity were measured accordingly.

It was found out that TSI result of 65-82, indicated that MSR was categorized as a eutrophic-hypereutrophic reservoir, with the dominance of Cyanophyta in 2011, and of

* E-mail: iclou@umac.mo.

Chlorophyta and Bacillariophyta in 2012. The PCR (including multiplex PCR) results showed that the techniques were successful for identifying cyanobacterial species and cyanotoxins-producing genotypes in pure cultures (or plasmid), mixed cultures, and water samples in MSR. While qPCR results were proved to be applied in quantifying the cell number of cyanobacteria, Microcystis spp. and C. raciborskii, as well as the gene copy nember of microcystin- and cylindrospermopsin-producing genotypes.

When the target species above 1 million cells/L, similar cell numbers estimated by microscopic counting and qPCR were obtained. Further quantification in water samples indicated that the ratio of cells number estimated by microscopy and by qPCR was 0.4-12.9 for cyanobacteria and 0.2-3.9 for C. raciborskii. However, Microcystis spp. was not observed by manual counting, while it can be detected at low levels by qPCR, suggesting that qPCR is more sensitive and accurate than microscopic counting.

On the aspect of cyanotoxins, there was a strong correspondence between the presence of the pks gene numbers and cylindrospermopsin concentrations (R2=0.95) determined by HPLC, while weak correlations were obtained between the mcys gene numbers and microcystin concentrations. Furthermore, the pks gene numbers were strongly related to Cylindrospermopsis (R2=0.88), cyanobacterial cell numbers (R2=0.96), total algae numbers (R2=0.95) and chlorophyll-a concentrations (R2=0.83), consistent with the dominant species of Cylindrospermopsis among the cyanobacteria existing in MSR. The water quality parameters NH4-N (R2=0.68) and pH (R2=0.89) were most highly correlated with the pks gene numbers.

INTRODUCTION

Cyanobacteria are responsible for freshwater algal blooms in eutrophic lakes or reservoirs. Macau Storage Reservoir (MSR), the main drinking water storage reservoir in Macau, has experienced algal blooms in recent years, with a high level of cyanobacteria, particularly *Microcystis* spp. and C. *raciborskii*, by microscopy. Recent study (Zhang et al. 2013) showed that MSR had a Tropic State Index (TSI) of 58–72 in 2010 and 64–82 in 2011. The TSI values, along with the detection of high levels of cyanobacteria in the summer season (of 40–200 x 10^6 cells/L), indicated that MSR is considered a eutrophic or hyper-eutrophic reservoir. Cyanobacterial blooms can deteriorate the water quality by reducing transparency, decreasing biodiversity, releasing taste and odor-causing compounds, and the most importantly producing cyanotoxins that pose a serious health hazard for humans (Paerl and Huisman, 2009). Microscopy-based surveys of the cyanobacteria indicated that the dominant species were *Microcystis* and *Cylindrospermopsis* (most of which are C. *raciborskii*). Detectable levels of microcystin and cylindrospermopsin were also measured. *Microcystis* spp. and *Cylindrospermospsis* spp. are two representative bloom-forming toxic genera (Chorus and Bartram 1999; Sivonen and Jones 1999; Duy et al. 2000). They are of great concern due to the potential release of toxic microcystin or cylindrospemopsin, which can disrupt liver cytoskeleton production by inhibiting protein phosphatases 1 and 2A (MacKintosh and Klumpp 1990), and thus are recognized as potential tumor promoters and carcinogens. However, there are very few studies that have quantified both microcystin- and cylindrospermopsin-producing genotypes simultaneously in freshwater reservoirs. The World Health Organization (WHO 1998) set the drinking water quality guideline of 1.0 µg /L as microcystin-LR equivalent (Falconer et al., 1999).

Cyanobacteria species have been traditionally identified on the basis of their microscopic morphology, physiology and staining characteristics, which are limited primarily by the inadequacy and inaccuracy, leading to misidentification due to their similarities. What's more, previous studies (Rapala et al. 1993; Orr and Jones 1998; Song et al. 1998; Wiedner et al. 2003; Rasmussen et al. 2008; Orr et al. 2010) have attempted to determine the relationship among the cyanobacteria, the corresponding cyanotoxin concentrations and the water parameters. However, no consistent conclusion can be drawn from the currently available data, because toxic and non-toxic strains often coexist in a bloom, and cannot be distinguished by conventional morphology-based identification methods (Kurmayer et al. 2002; Welker et al. 2004; Ouahid et al. 2005). Manual counting of cyanobacteria is also not accurate due to morphological similarity (Zhang et al. 2013). Quantification of those species was for a long time impedes, and even skilled and experienced technicians are occasionally unable to identify and enumerate the species in such environmental samples. In addition, water parameters that lead to a dominance or repression of toxin-producing genotypes versus nontoxic genotypes are not well understood. Therefore, identifying the conditions that lead to growth of potential toxin-producing strains and actual toxin synthesis in a specific water body are the first steps to improving water quality control (Humpage 2008).

For a couple of past decades, advances in molecular methods for evaluating microbial diversity in natural environments enable us to develop new techniques for identification and quantification of microorganisms. Early studies (Scholin et al., 1994) used the fluorescence in situ hybridization (FISH) approach coupled with microscopy or with flow cytometry to count cell number of strain-specific harmful algal species. Recently, PCR based on housekeeping genes, 16S/23S ribosomal RNA sequence, and rpoC1 database are now available for researchers to use rDNA-targeted hybridization for studying cyanobacteria. The phylogenetic methods have the following advantages compared to the traditional identification and quantification methods: i) Molecular methods are more accurate: unlike morphology, physiology and staining characteristics, conserved phylogenetic identity does not change over time or under different conditions. ii) The method can yield target-specific quantitative data. By targeting multiple cyanobacteria, multiplex PCR consisting of multiple primer sets in a single PCR mixture can be developed (Al-Tebrineh et al., 2012). These techniques have been subsequently applied for the detection and quantification of toxic genotypes, *mcy* genes and *cyr* gene of *Microcystis* spp. and C. *raciborskii*, respectively (Ha et al., 2009; Rasmussen et al., 2008), as they have been found to correlate with microcystin and cylindrospermopsin concentrations in pure cultures and environmental samples. PCR has been used for targeting cyanotoxin biosynthesis genes (Pearson and Neilan 2008) by designing primers that are highly sensitive and specific. Starting from the first detection of toxic *Microcystis* strains by Tillett et al. (2001), subsequent studies have developed multiplex PCRs targeting various *mcy* genes simultaneously (Ouahid and del Campo 2009; Valério et al. 2010), which have been proven reliable in predicting toxin production in water samples. PCR is inherently a qualitative procedure, so to quantify the number of genes of interest present in a sample, and thus gain an estimate of the number of organisms potentially able to synthesize toxin in a bloom, quantitative real-time PCR (qPCR) can be applied. qPCR has been used to quantify toxin-producing cyanobacteria by targeting *mcy*A (Furukawa et al. 2006), *mcy*B (Kurmayer et al. 2002), *mcy*C (Herry et al. 2008), *mcy*D (Rinta-Kanto et al. 2005), *mcy*E (Vaitomaa et al. 2003), *mcy*G (Ngwa et al. 2012), *mcy*J (Kim et al. 2010; Joung et al. 2011) and *pks* (Rasmussen et al. 2008; Moreira et al. 2011). Although qPCR does not provide information about actual cyanotoxin concentrations, but indicates only the potential risk of toxin synthesis, it is specific for detection of toxic strains of cyanobacteria, and is increasingly applied to monitoring the toxin-producing genotypes of cyanobacteria in water bodies (Pearson and Neilan 2008; Sivonen 2008; Kurmayer and Christiansen 2009).

This study aimed to develop and apply PCR (including mutiplex PCR) and real-time PCR to monitor simultaneously the spatial and temporal distribution of cyanobacteria, *Microcystis* spp. and *C. raciborskii* that are the dominant blooming causative species and characterize the spatial and temporal variation of both cylindrospermopsin- and microcystin- producing genotypes, and the corresponding cyanotoxin concentrations in MSR. Water samples and water parameters at three different locations with two sampling depths each were investigated monthly from September 2011 to August 2012. qPCR was applied to quantify the non-toxic and potentially toxic *Microcystis* and *Cylindrospermopsis* genotypes, along with changes in their population dynamics.

To our knowledge, this is the innovative study that integrating the multiplex PCR and real-time PCR for identifying and quantifying *Microcystis* spp. and C. *raciborskii* simultaneously in freshwater water reservoir. qPCR was applied to quantify the non-toxic and potentially toxic *Microcystis* and *Cylindrospermopsis* genotypes, along with changes in their population dynamics.

We also attempted to test if the water parameters affect the proportion or cell number of potentially toxic *Microcystis* or *Cylindrospermopsis* genotypes, and to relate the populations to water parameters.

MATERIALS AND METHODS

Sampling Sites

The study was carried out in MSR (22°12′12′′N, 113°33′′12′′E), Macau, China, which is located in the east part of Macau peninsula, is the largest reservoir in Macau, with a surface area of 0.35 km^2, an average depth of 4.5 m and the capacity is about 1.9 million m^3. It receives raw water from the Pearl River of China, and can provide water supply to the whole areas of Macau for about one week. It is particularly important as the temporary water source during the salty tide period when high salinity concentration is caused by intrusion of sea water to the water intake location. This region is characterized by a humid subtropical climate, with seasonal climate greatly influenced by the monsoons. Our recent study (Zhang et al., 2013) showed that MSR has the Tropic State Index (TSI) of 58-72 in 2010 and 64-82 in 2011, which is categorized as a eutrophic or hyper-eutrophic reservoir, and a high level of cyanobacteria of 40-200 millions cells/L was detected in summers. And the situation appears to be getting worse in recent years, with high phytoplankton abundance of *Microcystis* and *Cylindrospermopsis* (most of which are *C. raciborskii*), identified and detectable levels of microcystin and cylindrospermopsin.

During the study period, water samples were collected monthly in MSR from September 2011 to August 2012. Six sampling points were selected at three stations S1–S3 (Figure1) with two different water depths (0.5 m and 3.5 m below the water surface) each. Stations S1, S2 and S3 are located in the inlet, center and outlet of MSR, respectively.

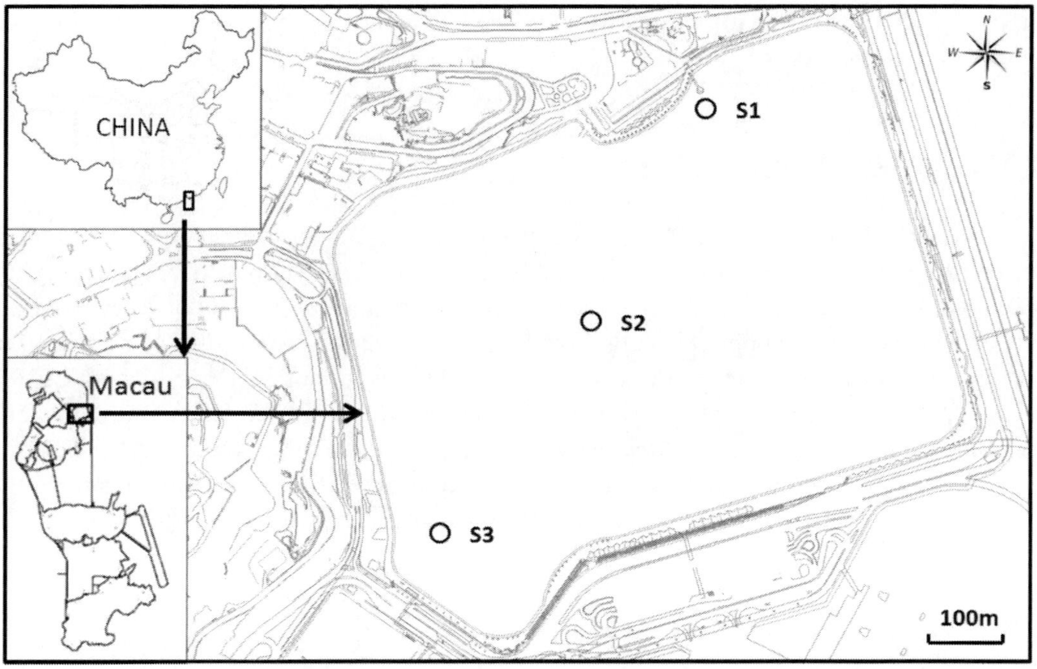

Figure 1. Layout of MSR and location of water sampling points.

P1 (P2), P3 (P4), and P5 (P6) are defined as the sampling points at 0.5 m (3.5 m) below the water surface at Stations S1, S2 and S3, respectively. Each sample of 2 L was kept at –80°C prior to analysis.

Cyanobacterial Strains and Culturing

The axenic strains of *Microcystis aeruginosa* (FACHB-905) and *C. raciborskii* (FACHB-1096) were obtained from the Freshwater Algae Culture Collection of the Institute of Hydrobiology, Wuhan, P.R. China. Strains were cultivated and grown as batch cultures in BG11 media (Sigma-Aldrich) at temperature of 25°C and under irradiance of 1000 lx. Both strains were harvested during the stationary phase for performing PCR and qPCR experiments in pure cultures and mixed cultures.

Enumeration of Cyanobacteria Using Microscopy

Pure cultures (10-100 μL) or well-mixed water samples (1 mL) were preserved by adding one drop of Lugol's iodine before sedimentation for 72 h. The sludge samples were stored at room temperature. The strains of cyanobacteria present in sedimentation chambers, including *Microcystis* and *Cylindrospermopsis*, were counted manually by inverted microscope using a Sedgwick-Rafter chamber from the method of McAlice, (1971). At least three transects per chamber were screened to enumerate cyanobacteria strain at ×100 magnification, according to morphological criteria. The total cyanobacteria were estimated as the sum of all the different species of cyanobacteria.

DNA and Plasmid Preparation

Densely grown pure culture (1 mL) or 175 mL of well-mixed water samples were taken and centrifuged in 3000 rpm for 10 min to obtain the cyanobacteria pellets. Considering the buoyant cells such as *Microcystis* do not settle easily at even at high speed of centrifugation, filtration followed by scrapping the cells from the filter paper was used. DNA extraction from the remaining pellets was carried out using QIAGen DNeasy Plant Mini Kit (Cat.No.69104, QIAGen, MD), according to the manufacturer's instructions. Extracted DNA was stored at -80°C. Since the pure culture of *C. raciborskii* purchased is a non-toxic strain with no *pks* gene detectable by PCR and no cylindrospermopsin measured by HPLC, a plasmid containing the synthesized *pks* gene was obtained from Invitrogen Trading (Shanghai) Co., Ltd.

PCR

Genomic DNA templates of pure cultures, mixed cultures and water samples were amplified by GeneAmp® PCR system 9700 (Applied Biosystems, CA) to demonstrate the presence of cyanobacteria, *Microcystis spp.* and *C. raciborskii* using specific primer sets (Table 1).

Table 1. Primers used for PCR and real-time PCR assays

Target	Primer	Sequence(5'-3')	DNA length	Ref
Cyanobacteria 16S rDNA	CYA359F	GGGGAATYTTCCGCAATGGG	470 bps	(Nubel et al., 1997)
	CYA781R(a)	GACTACTGGGGTATCTAATCCCATT		
	CYA781R(b)	GACTACAGGGGTATCTAATCCCTTT		
Microcystis 16S rDNA	MSR-S1f	TCAGGTTGCTTAACGACCTA	409 bps	(Otsuka et al., 1998)
	MSR-S2r	CTTTCACCAGGGTTCGCGAC		
C. raciborskii rpoC1	cyl2	GGCATTCCTAGTTATATTGCCATACTA	308 bps	(Rasmussen et al., 2008)
	cyl4	GCCCGTTTTTGTCCCTTTCGTGC		

Primers specificities with reference to the selected sequences in this study were verified using the BLAST program of the NCBI website (http://www.ncbi.nlm.nih.gov/tools/primer-blast/). Each PCR mixture contained 0.5 μL of DNA template solution (0.5 - 5 ng DNA/ 50 μL PCR), 5 μL of 10×PCR Buffer ($MgCl_2$ plus), 4 μL of the dNTP mixture, 0.5 μL (10 pmol) of each primer and 0.25 μL of *TaKaRa Taq*™ [20 mM Tris–HCl (pH 8.0), 100 mM KCl, 0.1 mM EDTA, 1 mM DTT, 0.5% Tween20, 0.5% Nonidet P40, 50% Glycerol solution]

(TaKaRa Biotechnology, China), and was adjusted to a final volume of 50 μL with sterile water (Sigma, USA).

The PCR was performed as follows: 95°C for 5 min, 30 cycles at 95°C for 30 s, 50°C - 60°C for 30 s and 72°C for 1 min, and a final extension step at 72°C for 7 min. Gel electrophoresis was run to confirm the presence of cyanobacteria, *Microcystis spp.* and *C. raciborskii* in the samples. All PCR products were separated on 2% agarose gels and observed on Molecular Imager ChemiDoc XRS System (Bio-Rad, USA) after staining with ethidium bromide for 20 min.

Multiplex PCR

In comparison to the single PCR, multiplex PCR has the advantage of targeting multiple genes rapidly and simultaneously from a number of cyanobacterial species and various cyanotoxin-producing genotypes in a single reaction. A multiplex PCR assay was developed that simultaneously targets the cylindrospermopsin-producing (*pks*) gene and *C.raciborskii*-specific (*rpoC1*) gene sequence, or mixed microcystin- producing (*mcy*) genes in water samples, using the corresponding primers (Table 1). Each multiplex PCR mixture contained 1 μL of DNA template solution (0.5–5 ng DNA/ 50 μL PCR), 25 μL of Multiplex PCR Mix 2, 0.5 μL (10 pmol) of each primer and 0.25 μL of Multiplex PCR Mix 1 (TaKaRa Biotechnology, China), and was adjusted to a final volume of 50 μL with sterile water (Sigma, USA).

The multiplex PCR was performed as follows: 94 °C for 1 min, 30 cycles at 94 °C for 30 s, 50 °C for 90 s and 72 °C for 90 s, and a final extension step at 72 °C for 10 min. All PCR products were separated on 4% agarose gels and observed on Molecular Imager ChemiDoc XRS System (Bio-Rad, USA) after staining with 6x loading dye for 20 min.

Real-Time PCR

Real-time PCR were performed in ABI 7500 Real-Time PCR system (Applied Biosystems, CA). All reactions were carried out in a total volume of 50 μL, containing 26 μL SYBR® *Premix Ex Taq*™ (Tli RNaseH Plus), including *TaKaRa Ex Taq* HS, dNTP Mixture, Mg^{2+}, Tli RNaseH, SYBR® Green I, plus ROX Reference Dye II (DRR420, TaKaRa Biotechnology, China), 1 μL forwards primers and 1 μL reverse primers (Table 1), 18 μL deionized water and 4 μL DNA templates. The thermal protocol for *Microcystis spp.* was that first warming 2 min at 50°C, and then preheating 10 min at 94 °C, followed by 40 cycles of 30 s at 94°C, 34 s at 50°C, then 1 min at 72 °C, and finally dissociated 15 s at 95°C, 1 min at 64 °C and 15 s at 99°C.

Fluorescence was measured at the end of each cycle at 72°C through channel F1 (530 nm) and a heating rate of 20°C s^{-1}. For cyanobacteria and *C. raciborskii* qPCR, thermal cycling steps were nearly the same as that for *Microcystis*, except for the annealing temperature that changed from 50°C to 60°C. All samples were amplified in triplicate. After qPCR amplification, fluorescent melting curve analysis was performed by gradually increasing the temperature from 72°C to 95°C at a rate of 0.1 s^{-1}. A correlation between the gene copy numbers and the threshold cycle number (Ct) (the cycle number at which the fluorescence exceeds the threshold) can be obtained.

Real-time PCR Standard Curves and Detection Limits

Standard curves and the detection limits for real-time PCR were established using 10-fold dilutions of the extracted pure culture DNA, with the concentrations ranging from 10^{13} to 10^4 cells/L. Corresponding primers (Table 1) were applied to the pure cultures of *Microcystis spp*., and *C. raciborskii*. Specific primers were used for targeting the pure culture of *C. raciborskii*, and the cell numbers were estimated as the *C. raciborskii* equivalent cells/L. The standard curves can be developed to relate Ct values to the cell numbers estimated by microscopy counting.

Water Parameters

The samples were analyzed for 14 abiotic parameters including water temperature (WT), secchi depth (SD), electrical conductivity (EC), pH, dissolved oxygen (DO), total nitrogen (TN), nitrate nitrogen (NO_3-N), nitrite nitrogen (NO_2-N), ammonia nitrogen (NH_4-N), total phosphorus (TP), orthophosphate phosphorus (PO_4-P), chlorophyll-*a* (Chl-*a*), microcystin and cylindrospermopsin concentrations. WT and SD were measured *in situ* with a mercury thermometer and a Secchi disk. The value of pH was determined in the laboratory with a pH meter (DKKTOA, HM-30R). Conductivity was measured with an EC meter (DKKTOA, CM-30R). DO, NH_4-N, NO_3-N, NO_2-N, TN, TP and PO_4^{3-} were measured according to the standard methods (APHA, 1995). Chl-*a* was determined using an UV-VIS Recording Spectrophotometer (SHIMADZU, UV-2401PC). Concentrations of microcystin and cylindrospermopsin were quantified by HPLC-MS/MS (Eaglesham et al. 1999).

In addition to abiotic parameters, the number of total algae and cyanobacterial cells were counted in the water samples. Well-mixed subsamples (1 mL) were preserved by adding one drop of Lugol's iodine before sedimentation for 72 h and storage at room temperature. The strains of total phytoplankton present in sedimentation chambers, including *Microcystis* and *Cylindrospermopsis*, were counted manually by inverted microscope using a Sedgwick-Rafter chamber from the method of McAlice (1971). At least three transects per chamber were screened to enumerate total algae and cyanobacterial strains at 100x magnification, according to morphological criteria.

RESULTS

PCR

The PCR amplification results (P1-P3, Figure 2) confirmed the feasibility of using cyanobacteria 16S rDNA, *C. raciborskii rpoC1 genes*, and *Microcystis* 16S rDNA for detection of the pure cultures of cyanobacteria, *C. raciborskii*, and *Microcystis spp.,* with corresponding bands of 470 base pairs (bps), 310 bps and 400 bps showed by gel electrophoresis, respectively. These results were consistent with the primers' DNA design lengths (Table 1). Before performing each qPCR, the gel electrophoresis results of real-time PCR products was used to confirm the presence of cyanobacteria, *C. raciborskii* and *Microcystis spp.* (T1-T6, Figure 2).

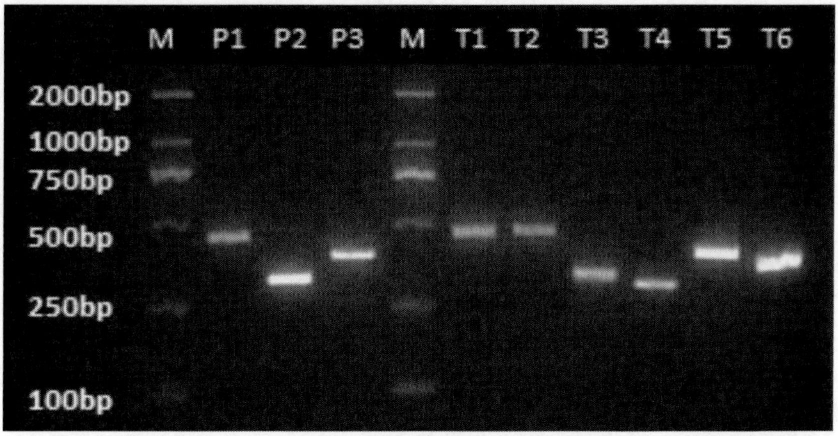

Figure 2. PCR amplification results of pure cultures tests for cyanobacteria 16S rDNA (P1), *C. raciborskii rpoC1* genes (P2), and *Microcystis* 16S rDNA (P3); verification of real-time PCR products of standard curve development for cyanobacteria 16S rDNA (T1 and T2), *C. raciborskii rpoC1* genes (T3 and T4) and *Microcystis* 16S rDNA (T5 and T6); M represented marker.

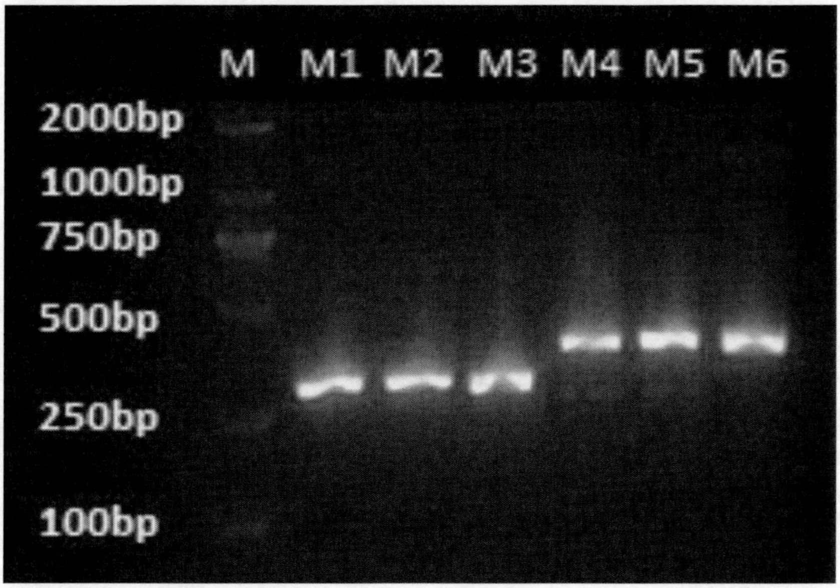

Figure 3. PCR amplification results of primers' specificities targeted for *C. raciborskii rpoC1* genes with *C.raciborskii/M.aeruginosa* = 9:1 (M1), *C.raciborskii/M.aeruginosa* = 1:1 (M2), and *C.raciborskii/M.aeruginosa* = 1:9 (M3); for *Microcystis* 16S rDNA with *M.aeruginosa/C.raciborskii* = 9:1 (M4), *M.aeruginosa/C.raciborskii*=1:1 (M5), and *M.aeruginosa/C.raciborskii* = 1:9 (M6); M represented marker. *C.raciborskii* primers were used for lanes M1-M3, and *Microcystis* primers were used for lane M4-M6.

Mixed cultures of *M. aeruginosa* and *C. raciborskii* were used for verifying specificity of the primer sets targeted for *C. raciborskii rpoC1* and *Microcystis* 16S rDNA. Pure cultures were mixed in three ratios, 9:1, 1:1 and 1:9 for both species. PCR results (Figure 3) confirmed the presence of *C. raciborskii rpoC1* (M1~M3) and *Microcystis* 16S rDNA (M4~M6) in mixture of both species, indicating the specificity of the primers used for mixed cultures. Comparison of the cells number estimated using microscopy and qPCR (Table2) showed that the difference was very small, with the ratio of around 0.94 to 1.16, further proving the selected primer sets were highly specific in mixed cultures conditions.

The PCR products in Figure 4 confirmed the presence of cyanobacteria, *C. raciborskii* and *Microcystis spp.* at the sampling point of 0.5 m from the water surface in Station S1 (P1). Similar results were also observed to those in other sampling points (not shown here). Moreover, multiplex technique was also showed to be feasible when applying to water samples in this study (Figure 5).

Multiplex PCR provides such advantages as reduction in assay time, labor and chemical cost. However, careful primer design (especially considering DNA oligomer melting temperature T_m), extensive optimization of reagents and condition were required to avoid non-specific primer-dimers interactions and to obtain a well-balanced set of amplicons (Schoske et al., 2003; Al-Tebrineh et al., 2012).

Table 2. Comparison of cell numbers of specific gene targets as estimated by counting and real-time PCR in pure culture and mixed cultures

	Primer target	Pure cultures strains used	Counting number (cells/L)	Real-time PCR results (cells/L)
Pure cultures	Cyanobacteria 16S rDNA	*C. raciborskii*	3.7×10^{11}	2.9×10^{11}
			1.8×10^{11}	1.5×10^{11}
	C. raciborskii rpoC1	*C. raciborskii*	2.8×10^{7}	5.4×10^{7}
			1.2×10^{8}	1.9×10^{8}
	Microcystis 16S rDNA	*M.aeruginosa*	3.7×10^{5}	1.9×10^{5}
			1.8×10^{5}	1.8×10^{5}
Mixed cultures	*C. raciborskii rpoC1*	*C. raciborskii* : *M.aeruginosa* (9:1)	1.1×10^{8}	1.1×10^{8}
		C. raciborskii : *M.aeruginosa* (1:1)	6.0×10^{7}	6.4×10^{7}
		C. raciborskii : *M.aeruginosa* (1:9)	1.2×10^{7}	1.2×10^{7}
	Microcystis 16S rDNA	*M.aeruginosa* : *C. raciborskii* (9:1)	3.3×10^{5}	3.0×10^{5}
		M.aeruginosa : *C. raciborskii* (1:1)	1.9×10^{5}	1.8×10^{5}
		M.aeruginosa : *C. raciborskii* (1:9)	3.7×10^{4}	3.2×10^{4}

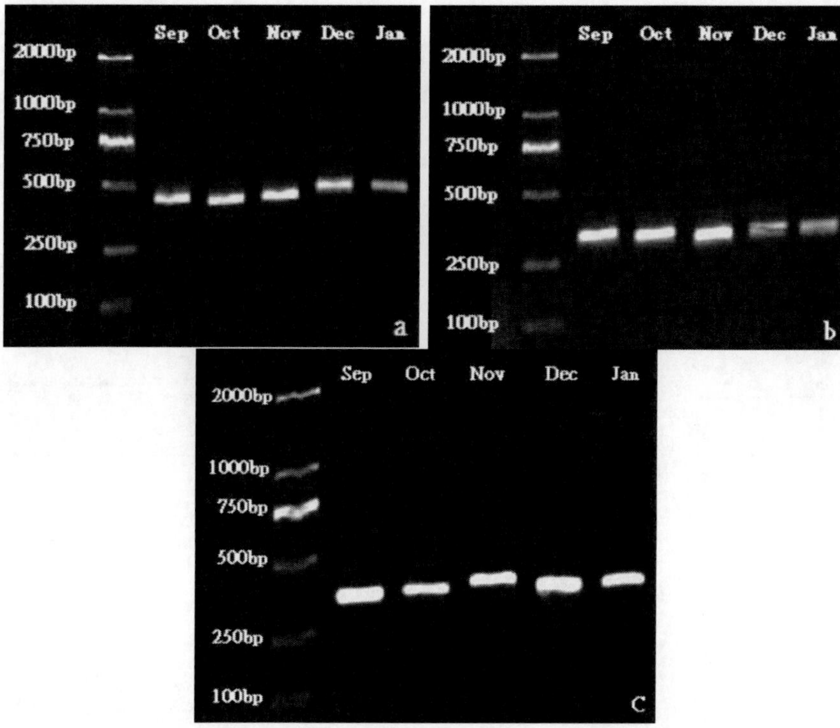

Figure 4. PCR amplification results of P1 water samples for cyanobacteria 16S rDNA (a), *C. raciborskii rpoC1* genes (b) and *Microcystis* 16S rDNA (c) from September to January.

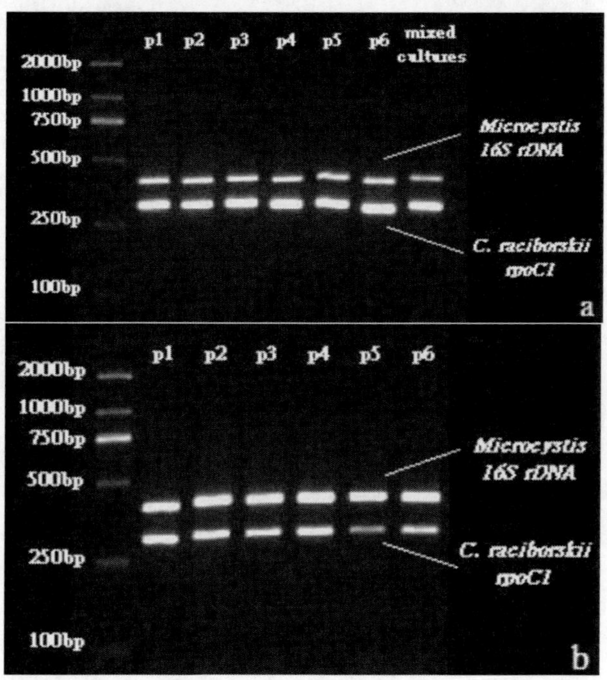

Figure 5. Multiplex PCR results of mixed cultures and water samples in November (a) and water samples in January (b) targeted for *C. raciborskii rpoC1* genes and *Microcystis* 16S rDNA.

qPCR

After applying the appropriate amount of amplified DNA with the corresponding primers for targeting specific class, species and strains, the copy number of the corresponding genes can be qualified with the Ct value. The standard curves were developed to relate the Ct values to the cell numbers estimated by microscopy. The change of fluorescence expressed as the delta Rn was showed in Figure 6 for cyanobacteria, *C. raciborskii* and *Microcystis spp.* using qPCR, with various dilution of initial DNA template.

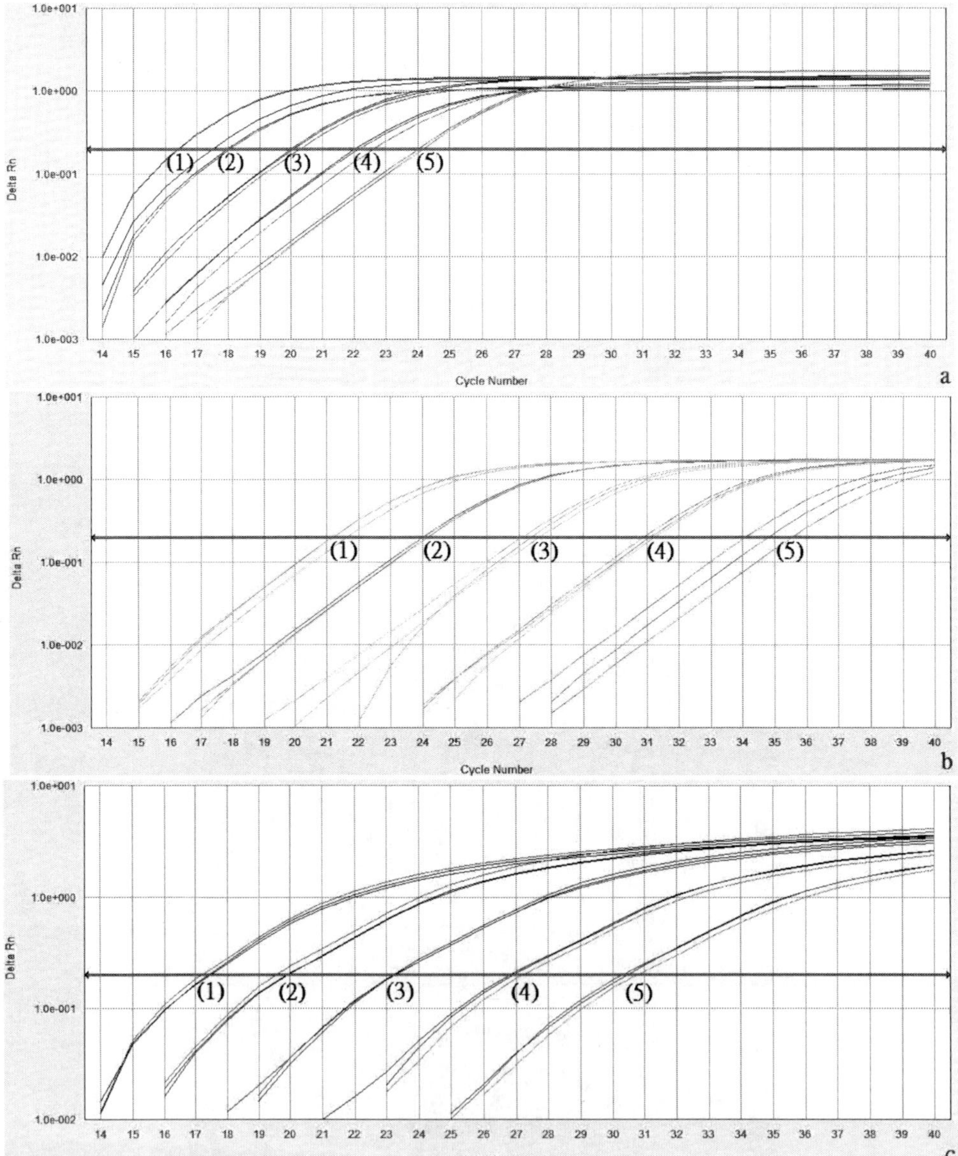

Figure 6. Change of delta Rn versus the cycle number (in triplicate) of 16S rDNA genes of cyanobacteria (a), *C. raciborskii* (b) and *Microcystis spp.* (c), with the initial DNA concentration by (1) no dilution, (2) 10 folds dilution, (3) 100 folds dilution, (4) 1000 folds dilution and (5) 10,000 folds dilution.

Table 3 showed the standard curves developed for cyanobacteria, *C. raciborskii* and *Microcystis* strains versus the Ct numbers, which have high correlation coefficients (>99%). For example, the regression equation was y=49.255-2.994x (R^2=0.993, p<0.0001) for rDNA of *C. raciborskii* , where y is the Ct and x is amount of initial DNA concentration as represented as the $\log_{10}$ (cell numbers). The cyanobacteria cell number estimated using the standard curves was expressed as *C. raciborskii* equivalent cells/L. The estimates of the slopes, 3.00 and 3.51, were similar to the theoretical value of 3.32 ($1/\log_{10}2$), which is an exact double values for each polymerization cycle (Larionov et al., 2005). To further test the reliability of these standard curves, two independent tests for each pure culture strain were conducted by comparing the cell numbers estimated using qPCR with those counted manually. Similar cell numbers with the population ratio of around 0.52 to 1.95 were obtained (Table 2), suggesting that the standard curves established in our study were able to estimate the cell number of cyanobacteria, *C. raciborskii* and *Microcystis spp.* in pure cultures. Furthermore, using *C. raciborskii* targeted by cyanobacteria primers are successful in quantifying the cyanobacteria abundance. However, it has to be careful when applying this quantification method to cyanobacteria, as different growth phases of cyanobacteria may affect the DNA copies of the cell number (ranging from 3 to 100) (Vaitomaa et al., 2003).

The sensitivity analysis in this study confirmed the high sensitivity of the qPCR using SYBR Green, in which it can be used to detect as low as several cells per milliliter of total cyanobacteria. By diluting the initial DNA amount to the appropriate folds, the sensitivity of qPCR is approximately 50 cells/mL of *C. raciborskii*, and around 140 cells/mL of *Microcystis* spp. Our reliable detection limits were similar to those obtained by Moreira et al., (2011), in which the detection for cyanobacteria achieved up to 11 cells/mL and for *C. raciborskii* can reached a limit of 258 cells/mL.

Using qPCR, the cell numbers of cyanobacteria, *Microcystis spp. and C. raciborskii* were estimated and showed that higher levels of cyanobacteria species were detected by qPCR than by manual counting, with the concentrations of approximately 1,290,000 - 210,150,000 cells/L (counting) versus 1,410,000 - 395,000,000 cells/L (qPCR) for cyanobacteria, and 820,000 - 84,000,000 cells/L (counting) versus 665,000 - 152,000,000 cells/L (qPCR) for *C. raciborskii* (Figure 7). It was noted that the cell numbers of cyanobacteria and *C. raciborskii* estimated by qPCR results were close to those estimated by manual counting, with the ratios of 0.4 - 12.9 and 0.2 - 3.9, respectively. All these data suggested that qPCR can be applied to water samples, for quantifying the abundances of cyanobacteria and *C. raciborskii*.

Table 3. Standard curve parameters from real-time PCR for the cyanobacteria, *C. raciborskii* and *Microcystis spp*

Target	Slope	y-intercept	r^2
Cyanobacteria 16S rDNA	-1.656	33.951	99.2%
C. raciborskii rpoC1	-2.994	49.255	99.3%
Microcystis 16S rDNA	-3.510	44.262	99.6%

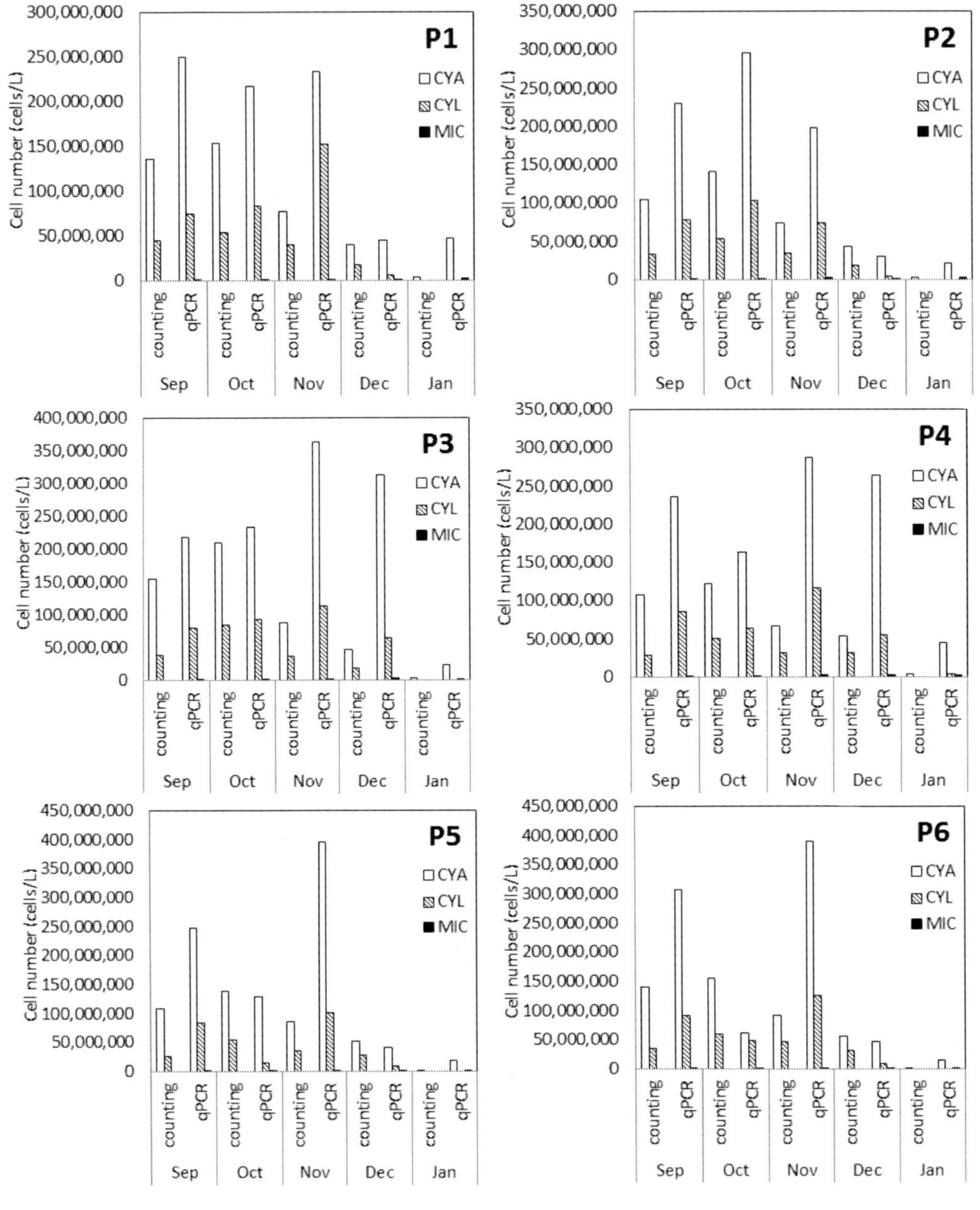

Figure 7. Comparison of qPCR results and cell counts of water samples for cyanobacteria (CYA), *Cylindrospermopsis* (CYL) and *Microcystis* (MIC) in each sampling point (P1~P6).

However, it was interesting to note that there is no *Microcystis* spp. observed in all of 30 water samples using microscopy due to the low number of *Microcystis* cells in the microscopic count, while a low level of 214,000 - 2,480,000 *Microcystis* cells/L were detected using qPCR.

These qPCR results were consistent with the PCR results shown above, implying that the method of qPCR is more sensitive and accurate than manual counting, in which one cell counted in the microscopic image after dilution was converted to 195,000 cells/L in the original water samples.

All these results revealed that the qPCR developed here can be used to quantify the *Microcystis* and *Cylindrospermopsis* when their concentrations were low, or even cannot be detected by traditional counting. Similar cell counts obtained by both methods when the *C. raciborskii* abundance in MSR exceeded around 1 million cells/L. However, substantial difference occurs when the organism concentration is low, for example lower than 1 million cells/L of *Microcystis* in this study. The cell counting using microscopy seems to underestimate the abundance of *Microcystis*, which was consistent with the findings by Tomioka et al., (2008), but contradicted the results by Artz et al., (2006) who found that dead cells in water samples can produce positive results by microscopy, leading to potential overestimation of the cell numbers. Thus the differences by both quantification methods should be carefully examined in the future. Furthermore, the relationship between cyanobacteria species abundance is not clearly understood by microscopy, due to the difficulty of inaccuracy measurement of cells abundance. It is believed that this difficulty can be overcome by qPCR. Thus, compared to traditional microscopic counting, qPCR should be more appropriate in monitoring the cyanobacteria abundance in MSR.

Considering the *C. raciborskii* was the dominant strains in MSR, the cyanobacteria quantified here was expressed as *C. raciborskii* equivalent cells/L that is based on the standard curve developed from the pure culture *C. raciborskii*, which needs to be carefully interpreted. Cyanobacteria frequently contain more than a single species in reservoirs, and the gene copy numbers of cyanobacteria 16S rDNA amplified as expression of different species are different, leading to the variations of cyanobacteria cells number estimated by qPCR.

Further statistical analysis using cluster analysis with at (Dlink/Dmax) < 20 (not shown here) indicated that there is no significant difference of cells number between each of two sampling points. Since *Cylindrospermopsis* is a thermophilic species, it can grow fast and be the dominant species during summer and autumn, but decrease in a large amount in low temperature seasons (Alster et al., 2010), which agreed with the qPCR percentage results shown in Figure 8.

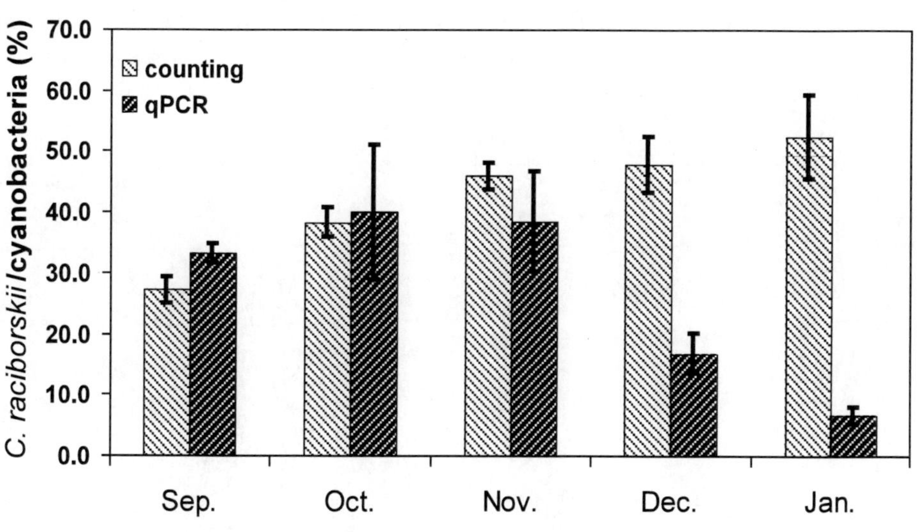

Figure 8. Comparison of the percentages of *C. raciborskii* over cyanobacteria from September 2011 to January 2012 by cell count and qPCR. The percentages were taken as the average over the samples and the error bars represent the standard deviation of the six sampling points.

The ratios of *C. raciborskii* over the total cyanobacteria estimated by both quantification methods (Figure 8) were showed to be similar from September to November, while to have much difference in December and January, which is due to the underestimation of cells number by microscopy MSR as described above. The ability to sensitively and accurately quantify the toxic strains presence in the water sample is the first step toward understanding the cyanobacteria dynamics and risks of algal bloom, though quantifying the abundance of potential toxin producers does not imply toxicity unless toxins, such as microcystin and cylindrospermopsin, are actually measured.

Furthermore, the quantification of C. *raciborskii rpoC1*, *Microcystis* 16S rDNA, *pks* and *mcy* genes in water samples can be performed by q-PCR. Standard curves were developed to relate the Ct values to gene copy number, which had strong correlations with an r^2 greater than 99%. Based on the standard curves, the *C. raciborskii rpoC1, Microcystis* 16S rDNA, *pks* and *mcy* genes in the water samples were quantified (Figure 9 and Figure 10). The spatial and temporal variation using comparison statistical analysis (data not shown) showed that 712 out of a total of 720 sampling points (with 6 different stations in a group), for each specific month, showed no significant differences among different sampling points, while all the sampling points (with 12 different months in a group), for each specific station, showed significant differences. The other eight sampling points that have significant difference in spatial analysis were *C. raciborskii rpoC1*, *mcyA, mcyB,* and *mcyC* in December 2011, *mcyA* in January 2012, *mcyJ* in May 2012, and *C. raciborskii rpoC1* in July 2012. These results were consistent with our previous findings that there is no significant difference of phytoplankton compositions or their corresponding abundances of different species including *C. raciborskii* and *Microcystis* spp. in Macau MSR (Zhang et al. 2014; Zhang et al. in press), leading to the similar genotypes copy number in different sampling points. The spatial similarity in the genotypes may be due to the small size of the reservoir, which easily make the water homogenous. We found that high concentrations of *pks* genes were measured in October 2011 (maximum 75,261 copies/μL at sampling point P2), which then dramatically decreased by December 2011 (to only 4 copies/μL at sampling point P1) (see Figure9a). The low concentration of *pks* genes was then maintained until the end of the study. A similar trend also occurred for *C. raciborskii rpoC1* genes. The variations in abundance of *mcy* and *Microcystis* 16S rDNA genes showed a more complicated pattern (Figure10).

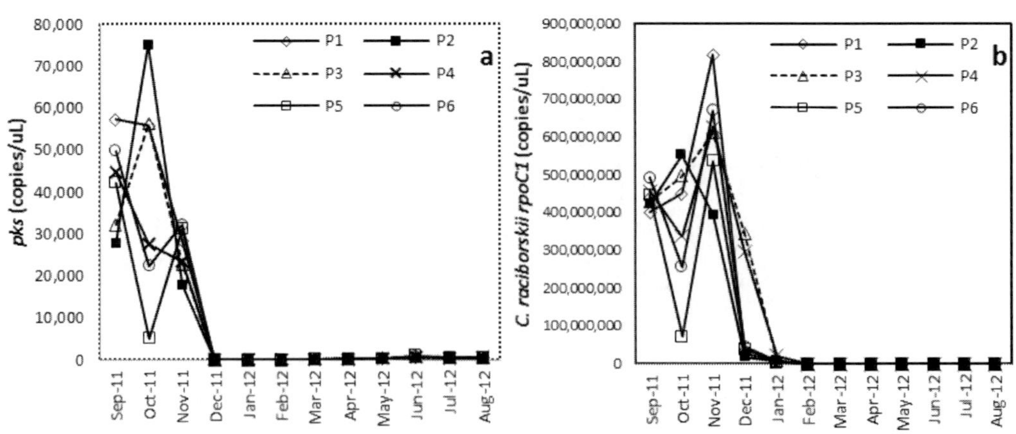

Figure 9. *pks* and C. *raciborskii rpoC1* gene abundances at different stations.

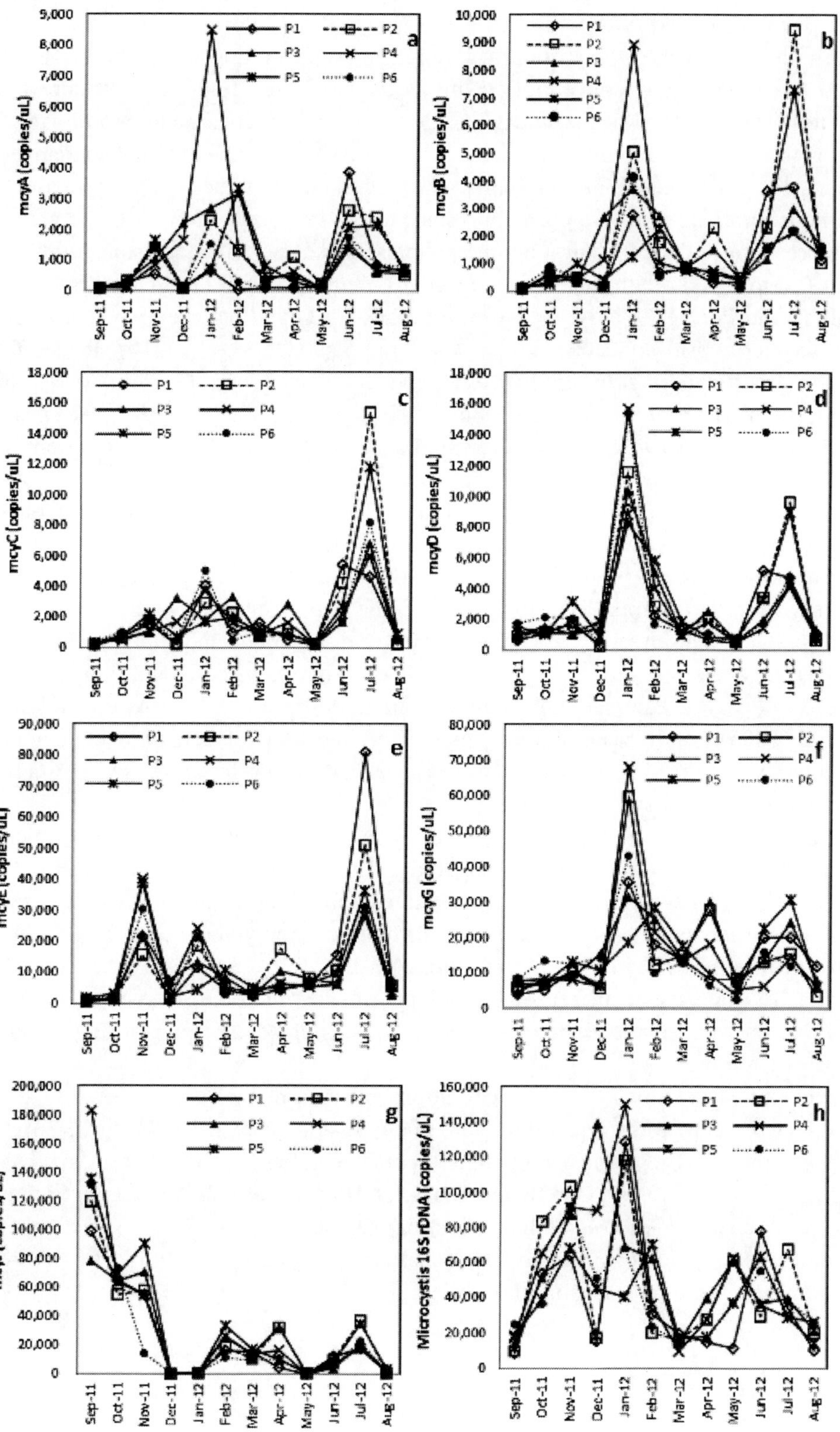

Figure 10. *mcyA/B/C/D/E/G/J* and *Microcystis* 16S rDNA gene abundances at different stations.

mcyA/B/C/D/E/G and *Microcystis* 16S rDNA genes had two peaks in January and June-July, while the *mcyJ* gene was present at its highest concentration in September but rapidly decreased to a very low concentration in December 2011 and then remained at low concentrations until the end of the study. Except for the *mcyJ* gene, the temporal variations of the *mcy* genes were essentially consistent with the dynamic change in abundance of the *Microcystis* 16S rDNA gene. Further analyses were conducted in the following sections.

The high correlation of *pks* genes abundance to *C. raciborskii rpoC1* gene copy numbers and cylindrospermopsin concentration in the present study were consistent with previous findings of a strong correlation between *pks* genes abundance and *C. raciborskii rpoC1* gene copy numbers in Portugal Vela Lake (Moreira et al. 2011) and between *pks* gene abundance and cylindrospermopsin concentration in water samples (Orr et al. 2010). In addition, the highly linear correlation between the *pks* gene abundance and *Cylindrospermopsis* cell number or *C. raciborskii rpoC1* gene abundance suggested that the *pks* gene was mostly released by *Cylindrospermopsis* cells that could also be estimated by quantification of the *C. raciborskii rpoC1* gene.

The weak correlation between *mcy* gene abundances and *Microcystis* cell counts ($R^2 < 0.4$) can be explained by two main factors. The *mcy* genes can be produced by *Microcystis* spp., but also by other cyanobacterial species, so the quantification of this gene may not be specific to *Microcystis* spp. The presence of the *mcy* gene in natural populations of *Planktothrix* (Christiansen et al. 2003 and Kurmayer et al. 2004) and *Anabaena* (Rouhiainen et al. 2004) has been reported. Some *Microcystis* cells may also carry multiple copies of the *Microcystis* 16S rDNA gene and *mcyD* gene from cell division and DNA replication (Rinta-Kanto et al. 2005). Such metabolic activities affect the ratio of cell number to gene copy number directly, resulting in unstable and weak correlations between *Microcystis* cell counts and *mcy* gene copy numbers.

Similarly, weak correlations between *mcy* gene abundances and microcystin concentrations were also observed in our study. This result may be more complicated to explain, as it involves the mechanisms of gene expression, cooperation and competition among different *mcy* genes during the course of microcystin synthesis. Each *mcy* gene plays an important role in synthesizing microcystin, and destroy of *mcy* A, B, D, E (Tilett et al., 2000), or F (Nishizawa et al., 1999) would inhibit the microcystin synthesis. Thus it is probably that only existence of the complete *mcy* genes, instead of part of them, could have microcystin synthesis.

The complicated environmental conditions also affect the *mcy* gene expression, cooperation and competition, leading to more one type of gene over another in specific genus. Furthermore, the *mcy* gene cluster contains peptide synthetases, polyketide synthases and tailoring enzymes that are encoded by ten genes for *Microcystis* and *Anabaena*, or nine genes for *Planktothrix* and *Nodularia* (Tillett et al. 2000). Since the *mcy* genes and microcystin could come from various species with uncertain quantitative relations, the correlations between *mcy* genes and microcystin may vary.

Multiplex PCR

Multiplex PCR was performed for simultaneous detection of multiple target genes, and the results are shown in Figure 11. The product of which confirmed the presence of cyanobacteria, *C. raciborskii* and *Microcystis spp.* at the sampling point of 0.5 m from the water surface in Station S1 (P1). These results confirmed the successful application of PCR for the simultaneous detection of *pks* and *C. raciborskii rpoC1* genes, and *mcy* and *Microcystis* 16S rDNA genes in water samples.

Multiplex PCR provides such advantages as reduction in assay time, labor and chemical cost. However, careful primer design (especially considering DNA oligomer melting temperature T_m), extensive optimization of reagents and condition were required to avoid non-specific primer-dimers interactions and to obtain a well-balanced set of amplicons (Schoske et al., 2003; Al-Tebrineh et al., 2012).

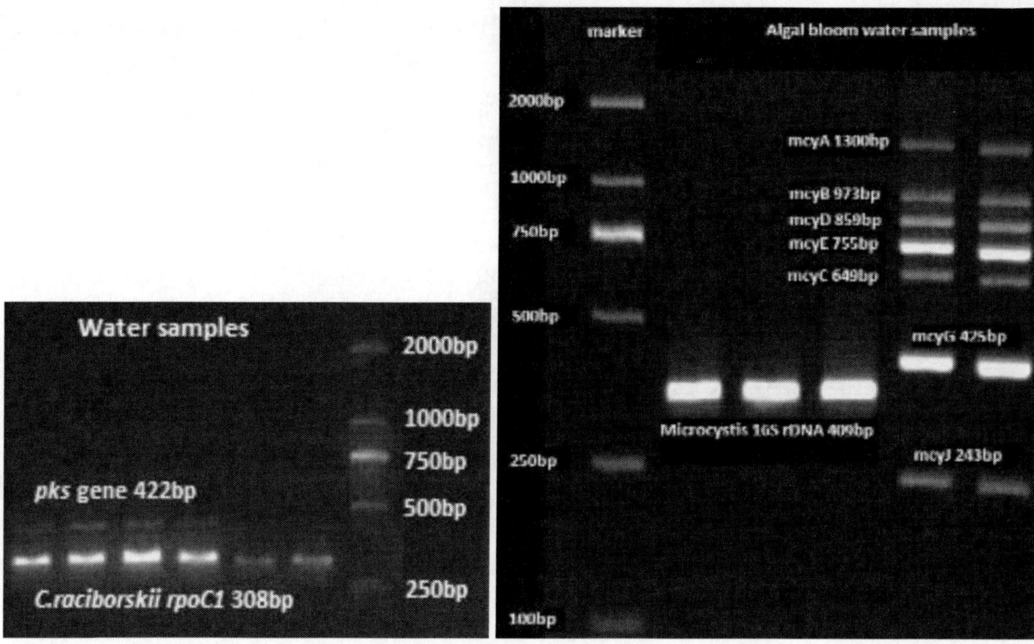

Figure 11. Multiplex PCR results of *pks/C. raciborskii rpoC1* and *mcyA-J/Microcystis* 16S rDNA in water samples.

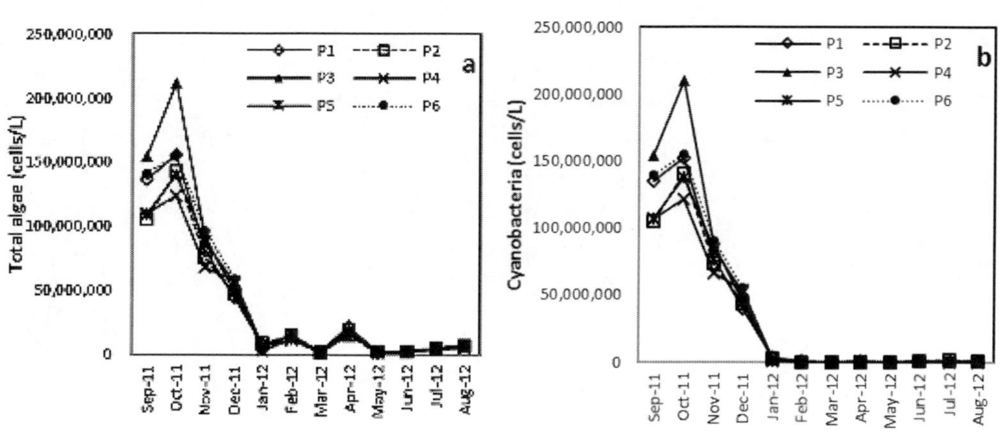

Figure 12. Variations in total algae and cyanobacterial cell counts at 6 points.

Total algae (also called phytoplankton) and cyanobacteria were counted and the results are shown in Figure 12.

Cyanobacteria were the dominant genus, accounting for more than 90% of total algae at most sampling timepoints. Algal and cyano-bacterial abundances showed similar patterns to the *C. raciborskii rpoC1* gene abundance, *pks* gene abundance and cylindrospermopsin concentrations, in that high abundances occurred in the fall 2011 samples, followed by a rapid decrease and low abundances until the end of the study.

Due to the short hydraulic retention time of MSR, it was believed that the communities and compositions of cyanobacteria are affected by the hydrodynamic parameters (such as water depth and rainfall) and physicochemical parameters (such as temperature, pH, nitrogen and phosphorus concentration, light intensity, and iron concentration) (Ha et al., 2009; Rinta-Kanto et al., 2005). In addition to cyanobacteria species, the cyanotoxins, microcystins and cylindrospermopsins are necessary to monitored routinely to meet the water quality regulations. Cylindrospermopsin and microcystin concentrations are shown in Figure 13. Both cyanotoxins showed significant temporal differences, but little spatial difference among the six sampling points.

Hence, further studies will be focused on quantifying different compositions of the cyanobacteria species, measuring the cyanotoxins and environmental parameters in the reservoir, thus relating the environmental parameters to the abundance of different cyanobacteria species populations and cyanotoxins levels.

The number of water bodies affected by cyano-bacterial blooms has been increasing worldwide. In MSR *Microcystis spp.* was previously the dominant bloom-forming cyanobacteria, while starting from 2011, *C. raciborskii* turns to be the main species. The reasons causing the shift of dominant species are not understood completely, probably because MSR is the pumped reservoir, in which the water quality including the physical, chemical water parameters and cyanobacteria community, as well as their interaction, are directly affected by the influents. For examples, *C. raciborskii* is generally thought of as a tropical species due to its affinity for warm water temperatures (25-30°C) (Pidasak, 1997), which are common in Macau. Unfortunately, to our knowledge, there is no further available reference for deeply explaining the relationship of the species compositions and water quality parameters.

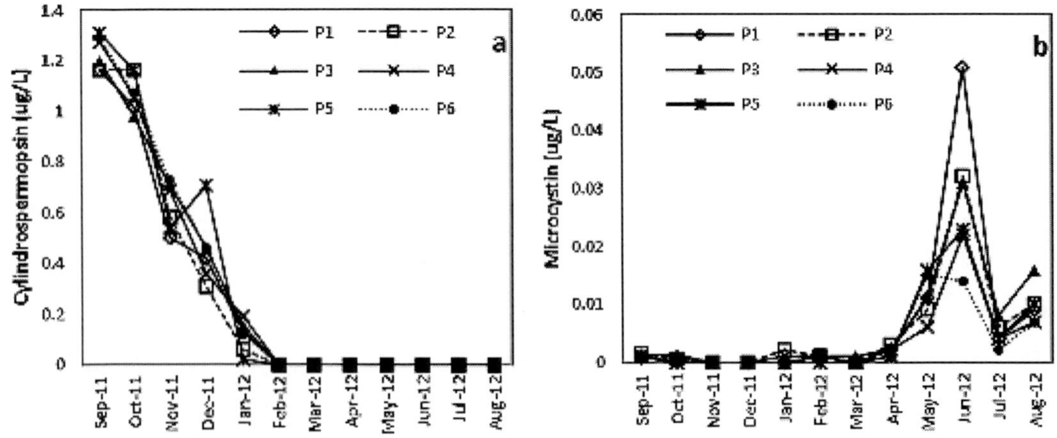

Figure 13. Concentrations of cylindrospermopsin and microcystin at 6 points.

It is thus of critical importance for water monitoring utilities to rapidly detect and quantify such potentially toxic cyanobacteria for the early detection and management of blooms in the reservoir using innovative techniques, as using the traditional microscopy only has some limitations, for example *Microcystis* strains usually form scum, which is difficult for enumeration, and *C. raciborskii* has many similarities of morphotypes such as *Anabaenopsis*, *Raphidiopsis*, and *Cylindrospermum* (Scholin et al., 1994), and it has been reported that until now, keys (morphological attributes) have not been developed for *C. raciborskii*, as a distinct species.

pH, NH_4-N and Chl-*a* are shown in Figure 14a–c and show similar patterns to those observed for the *pks* genes as described above. As the concentrations of TN, TP, and the N:P ratio are the common factors affecting the growth of cyanobacteria, their values were measured or calculated and are displayed in Figure 14d. TN concentrations ranged from 0.53 to 1.43 mg/L, with three peaks in September, Nov–Jan and March, while TP had only two peaks in Nov–Dec and Jul–Aug with the maximum value around 0.60 mg/L. The N:P ratio fluctuated in the range of approximately 1 to 4.

Chl-*a* concentration was significantly correlated with *pks* gene abundance. Chl-*a* is conventionally used as an indicator to assess the abundance of phytoplankton in water utilities. Chl-*a* therefore increases with an increase in the abundance of phytoplankton, including cyanobacteria.

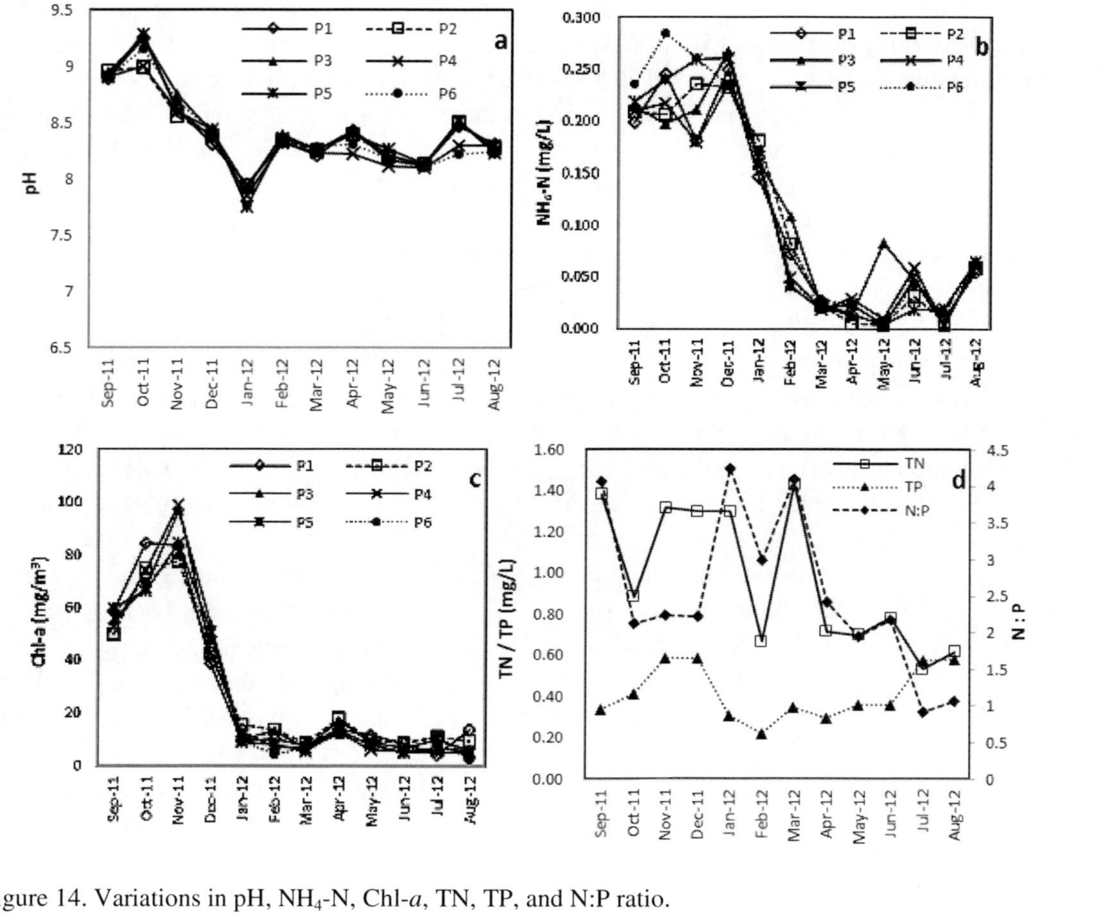

Figure 14. Variations in pH, NH_4-N, Chl-*a*, TN, TP, and N:P ratio.

In this study, cyanobacteria accounted for more than 90% of the phytoplankton abundance, and among the cyanobacteria present, *Cylindrospermopsis*, the cylindrospermopsin-producing species, was dominant. Thus the *pks* gene abundance would be strongly correlated with the Chl-*a* concentration.

In other reservoirs it has been found that TP concentration is the most important factor affecting the potential for cyanobacterial blooms, however in MSR the nitrogen source is the critical factor. In general, the growth rate of cyanobacteria was low in the absence of a fixed-nitrogen source and high in the presence of NH_4-N as the nitrogen source (Saker and Neilan 2001). NH_4-N can also play an indispensable role in phytoplankton metabolic activity, such as widening of cells and increasing biomass (Kohl et al. 1985). Therefore, higher NH_4-N concentrations increased the growth of *Cylindrospermopsis* leading to higher abundance of *pks* genes. This resulted in a strong correlation between NH_4-N concentrations and *pks* gene abundance. The recent research outcomes (Dolman et al., 2012) showed that *C. raciborskii* often reach their highest biovolumes with high N relative to P concentration, which matched our results that the highest C. *raciborskii rpoC1* gene copy number detected, with the high value of N:P ratio of 1.6 (Figure 14d) in November of 2011. It was also observed that in the summer of 2012, the population of C. *raciborskii* and *Microcystis* spp. changed in spite of high N:P ratio, which is probably due to the much low TN and TP concentrations. This finding agreed on the conclusion from Dolman et al. (2012) that cyanobacteria should not be treated as a single group when considering the potential effects of changes in nutrient loading on the phytoplankton community structure. However, different from the Dolman et al. (2012) study, the cylindrospermopsin is strongly related to the C. *raciborskii*. Furthermore, the two periods of low N: P ratio observed in Oct–Dec and July, supported the previous findings (Schindler, 1977) that high phosphorus concentrations and a low N: P ratio tended to favor the growth of cyanobacteria that are capable of fixing atmospheric nitrogen (the former condition favors *Cylindrospermopsis* and the latter favors *Microcystis*).

In this study, high cylindrospermopsin concentrations and *pks* gene abundances were accompanied by high pH ($\geq$ 8.5), which was consistent with previous findings (Westrick 2008) that cylindrospermopsin can be degraded effectively at pH values between 6.0 and 8.0. In contrast, Senogles et al. (2001) found that high pH (= 9) resulted in the best degradation rate of cylindrospermopsin. The contradictory results probably occur due to the different charge on ion components present in the water environment affecting toxin degradation, as low (or high) pH particles would have a positive (or negative) charge, and the overall charge on the components surface would affect the attraction and repulsion between cyanotoxins and catalytic materials, thus influencing the degradation rate of cylindrospermopsin directly (Lindner et al. 1995).

Except for *mcyJ*, all the other *mcy* gene abundances were strongly positively correlated with each other. The correlations between the *mcy* gene abundances and *Microcystis* cell counts, *Microcystis* 16S rDNA gene abundances and microcystin concentrations were weak, implying that *Microcystis* may not contribute to the *mcy* genes quantified in the water samples. Water parameters were also not strongly correlated with the *mcy* gene abundance, except for *mcyJ* which was highly correlated with total algae (R^2 = 0.83) and cyanobacteria (R^2 = 0.82), pH (R^2 = 0.82), NH_4-N (R^2 = 0.52) and Chl-*a* (R^2 = 0.68). Furthermore, it is surprising to observe that high positive correlation between *mcyJ* and cylindrsopermopsin (R^2 = 0.87), *Cylindrospermopsis* (R^2 = 0.70), C. *Raciborskii* (R^2 = 0.70) and *pks* gene (R^2= 0.92).

The correlations among the water parameters, *pks* gene copy number and *mcy* genes copy number were shown in Table 4.

These results showed that the abundance of *pks* genes was strongly correlated to the total algae, total cyanobacteria, Cylindrospermopsis cell numbers, C. raciborskii *rpoC1* gene copy numbers and cylindrospermopsin concentrations, with correlation coefficients of 0.95, 0.96, 0.89, 0.88 and 0.95, respectively.

The current study has demonstrated the validity of PCR use for the detection of cyanobacterial strains in eutrophic water, with more accurate and sensitive detection results shown above.

Table 4. Correlations of water parameters and *pks*/*mcy* genes ($p < 0.01$)

	pks	*mcyA*	*mcyB*	*mcyC*	*mcyD*	*mcyE*	*mcyG*	*mcyJ*
Temperature (°C)	0.22	-0.41	-0.12	0.07	-0.38	0.14	-0.54	0.22
SD (m)	-0.49	0.16	0.43	0.31	0.38	0.15	0.52	-0.28
Conductivity (μs/cm)	-0.43	0.10	-0.07	-0.28	0.16	-0.34	0.40	-0.37
pH	0.89	-0.60	-0.51	-0.22	-0.48	-0.13	-0.58	0.82
DO (mg/L)	0.23	-0.51	-0.17	0.19	-0.42	0.11	-0.42	0.21
TN (mg/L)	0.35	-0.02	-0.26	-0.34	0.05	-0.21	0.11	0.34
TP (mg/L)	0.06	-0.14	0.06	0.25	-0.19	0.45	-0.34	-0.05
NO_3-N (mg/L)	-0.56	0.27	0.03	-0.13	0.15	-0.26	0.39	-0.45
NO_2-N (mg/L)	-0.54	0.20	0.24	0.11	0.24	0.00	0.29	-0.51
NH_4-N (mg/L)	0.68	-0.05	-0.29	-0.32	-0.02	-0.18	-0.09	0.52
PO_4^{3-} (mg/L)	-0.02	-0.11	-0.21	-0.20	-0.10	-0.22	0.08	0.20
Chl-*a* (mg/m^3)	0.83	-0.30	-0.47	-0.29	-0.29	0.00	-0.37	0.68
Microcystin (μg/L)	-0.29	0.30	0.11	0.12	-0.09	-0.03	-0.13	-0.31
Cylindrospermopsin (μg/L)	0.95	-0.39	-0.46	-0.33	-0.26	-0.23	-0.37	0.87
Total algae (cells/L)	0.95	-0.42	-0.48	-0.34	-0.31	-0.22	-0.40	0.83
Cyanobacteria (cells/L)	0.96	-0.41	-0.47	-0.33	-0.30	-0.21	-0.41	0.82
Cylindrospermopsis (cells/L)	0.89	-0.36	-0.46	-0.31	-0.30	-0.15	-0.39	0.70
Microcystis (cells/L)	-0.38	0.34	0.35	0.37	0.01	0.19	-0.06	-0.33
C.raciborskii rpoC1 (copies/μL)	0.88	-0.27	-0.46	-0.27	-0.27	0.06	-0.37	0.79
pks (copies/μL)	1.00	-0.39	-0.45	-0.30	-0.27	-0.15	-0.39	0.92
Microcystis 16S rDNA (copies/μL)		0.62	0.35	0.18	0.57	0.35	0.52	-0.18
mcyA (copies/μL)		1.00	0.73	0.49	0.80	0.37	0.80	-0.37
mcyB (copies/μL)			1.00	0.86	0.87	0.66	0.78	-0.37
mcyC (copies/μL)				1.00	0.59	0.85	0.46	-0.17
mcyD (copies/μL)					1.00	0.47	0.95	-0.23
mcyE (copies/μL)						1.00	0.32	-0.06
mcyG (copies/μL)							1.00	-0.34
mcyJ (copies/μL)								1.00

Compared with PCR technique, traditional microscopy has the advantages of being cheap, simple, and able to identify the presence of cyanobacteria. However, it is time-consuming, and inability to distinguish the morphologically similar species. PCR, on the other hand, is specific, sensitive, and reproducible with the ability for adaption to many data measurement at the same time. The disadvantage of PCR is that it requires the knowledge of designing primers using gene sequences the database, for targeting specific species. Sterile environment should be maintained during the PCR operation, to avoid any contamination leading to amplification of non-specific products, which may affect the PCR results (Felske and Osborn, 2005). Besides, water samples often contain complicated mixtures of organic matter that may inhibit PCR. In spite of these disadvantages, the molecular methods are still undoubtedly useful for early effective detection of cyanobacteria (Humbert et al., 2010). As PCR is only qualitative, the application of qPCR can provide a quantitative option for the assessment of cyanobacterial community, which correlates the gene copy number measured to the cells number. Philip et al., (2010) pointed out several possible reasons could result in poor correlation between qPCR results and cell counting, such as failure to sediment negatively buoyant cells during centrifugation prior to DNA extraction; natural variability in gene cell quotas between different species and strains including polyploidy; genetic variation within the target regions of the primers at both strain and species level; sub-quantitative recovery of cells and DNA from samples, and over or underestimation of cell concentrations using traditional microscopy methods. Based on the qPCR results with the manual count, the integration of traditional microscopy with the PCR for routine monitoring programs for cyanobacteria will be probably developed in the future, with an early signal of dominant species identified and specific primers to the dominant species applied to quantify their concentrations in the reservoir for monitoring afterward.

CONCLUSION

MSR is experiencing the algal bloom problem these years, with the dominant species of *Microcystis spp.* and *C. raciborskii*. Traditional microscopic counting is currently being used in monitoring the cyanobacteria species, which is limited to inaccuracy and time-consuming. To overcome these limitations, PCR and qPCR were developed in this study. Our results showed that the techniques were proved to be successful for identification and quantification of cyanobacteria, *Microcystis spp.* and *C. raciborskii* in pure cultures, mixed cultures and water samples, suggesting that they are the promising technologies that can be used to replace the microscopic counting method, particularly for early detection of algal blooms formation. Besides, one finding of our study is developing a multiplex PCR that has the potential to address if the two or more dominant species are present in a certain water samples in a single PCR assay, which benefit the targeted users of the reservoir for further research and public health in a rapid and accurate approach. The quantification in water samples indicated that similar cyanobacteria and *C. raciborskii* cells number estimated by microscopy and by qPCR were obtained, with the ratios of 0.4 - 12.9 and 0.2 - 3.9, respectively. However, *Microcystis* spp. was not observed by manual counting, while it can be detected at low levels by qPCR, indicating qPCR is proved to be more sensitive and accurate than microscopic counting. Therefore, the molecular approaches offer new tools to address many remaining questions in fundamental and applied cyanobacterial ecology. In this study, the cylindrospermopsin synthetase gene (*pks*) and microcystin synthetase genes (*mcyA, mcyB, mcyC, mcyD, mcyE, mcyG and mcyJ*) were used to identify and quantify cylindrospermopsin- and microcystin-producing genotypes and gene copy numbers in MSR. Our results showed that high concentrations of cylindrospermopsin and low concentrations of microcystin were measured during the study period. There is a strong correlation between pks gene abundance and cylindrospermopsin concentrations, while there was only a weak correlation between *mcy* gene abundance and microcystin concentrations. It was also found that *pks* gene abundance was strongly related to Cylindrospermopsis, cyanobacterial cell numbers, total algae numbers and chlorophyll-a concentration. In addition, NH_4-N and pH were the water quality parameters most highly correlated with the pks gene abundance. And in sum, considering the advantages of PCR and qPCR over microscopy, the techniques, in addition to microscopic counting, can be potentially used for routine water quality monitoring in such risk water bodies as MSR.

ACKNOWLEDGMENT

We thank Larrisa Lei, the Master student in the Institute of Chinese Medical Science, for assistance with PCR and gel electrophoresis studies, the technical staff at Macao Water Supply Co. Ltd. for water sampling, and Xi Chen, undergraduate student in the Department of Civil and Environmental Engineering at University of Macau, for helping with DNA extraction and water parameter measurement.

The research project was supported by the Fundo para o Desenvolvimento das Ciências e da Tecnologia (FDCT) (grant# 069/2014/A2) and the Research Committee of University of Macau.

REFERENCES

Alster A., Kaplan-Levy R. N., Sukenik A., Zohary T. (2010). Morphology and phylogeny of a non-toxic invasive *Cylindrospermopsis raciborskii* from a Mediterranean Lake. *Hydrobiologia,* 639, 115-128.

Al-Tebrineh J., Pearson L. A., Yasar S. A., Neilan B. A. (2012). A multiplex qPCR targeting hepato- and neurotoxigenic cyanobacteria of global significance. *Harmful algae*, 15, 19-25.

Artz R. R. E., Avery L. M., Jones D. L., Killham K. (2006). Potential pitfalls in the quantitative molecular detection of Escherichia coli O157:H7 in environmental matrices. *Canadian Journal Microbiology*, 52(5), 482-488.

Chorus, I., and Bartram, J. (1999). Toxic Cyanobacteria in Water. A Guide to their Public Health Consequences, Monitoring and Management. London: EandFN Spon.

Christiansen, G., Fastner, J., Erhard, M., Börner, T., and Dittmann, E. (2003). Microcystin biosynthesis in Planktothrix: genes, evolution and manipulation. *Journal of Bacteriology*, 185, 564-572.

Conradie, K. R., and Barnard, S. (2012). The dynamics of toxic *Microcystis* strains and microcystin production in two hypertrofic South African reservoirs. *Harmful Algae*, 20, 1-10.

Davis, T. W., Berry, D. L., Boyer, G. L., and Gobler, C. J. (2009). The effects of temperature and nutrients on the growth and dynamics of toxic and non-toxic strains of *Microcystis during cyanobacterial blooms. Harmful Algae*, 8, 715-725.

Dittmann, E., Neilan, B. A., Erhard, M., von Döhren, H., and Börner, T. (1997). Insertional mutagenesis of a peptide synthetase gene that is responsible for hepatotoxin production in the cyanobacterium *Microcystis* aeruginosa PCC 7806. *Molecular Microbiology*, 26, 779-787.

Dolman A. M., Rücker J., Pick F. R., Fastner J., Rohrlack T., Mischke U., Wiedner C. (2012), Cyanobacteria and cyanotoxins: the influence of nitrogen versus phosphorus. *PLoS One*, 7, e38757.

Duy, T. N., Lam, P. K., Shaw, G. R., and Connell, D. W. (2000). Toxicology and risk assessment of freshwater cyanobacterial (blue-green algal) toxins in water. *Reviews of Environmental Contamination and Toxicology*, 163, 113-185.

Eaglesham, G. K., Norris, R. L., Shaw, G. R., Smith, M. J., Chiswell, R. K., Davis, B. C., Neville, G. R., Seawright, A. A., Moore, M. R. (1999). Use of HPLC-MS/MS to monitor cylindrospermopsin, a blue-green algal toxin, for public health purposes. *Environmental Toxicology*, 14, 151-154.

Falconer I., Bartram J., Chorus I., Kuiper-Goodman T., Utkilen H., Burch M., Codd G. A. (1999). Safe levels and safe practices, in Chorus I., Bartram J. (Ed.), Toxic Cyanobacteria in Water, A Guide to Their Public Health Consequences, Monitoring and Management (pp. 161-182). London: Spon Press.

Felske A., and Osborn A. M. (2005). DNA fingerprinting of microbial communities, in Osborn A. M., Smith C. J. (Ed.), Molecular Microbial Ecology (pp. 65-90). New York: Taylor and Francis.

Fergusson, K. M., and Saint, C. P. (2003). Multiplex PCR Assay for *Cylindrospermopsis raciborskii* and cylindrospermopsin-producing cyanobacteria. *Environmental Toxicology*, 18, 120-125.

Furukawa, K., Noda, N., Tsuneda, S., Saito, T., Itayama, T., and Inamori, Y. (2006). Highly sensitive real-time assay for quantification of toxic cyanobacteria based on microcystin synthetase A gene. *Journal of Bioscience and Bioengineering*, 102, 90-96.

Giglio S., Monis P. T., Saint C. P. (2003). Demonstration of preferential binding of SYBR Green I to specific DNA fragments in real-time multiplex PCR. *Nucleic Acids Research,* 31(22), e136.

Ha J. H., Hidaka T., Tsuno H. (2009). Quantification of toxic *Microcystis* and evaluation of its dominance ratio in blooms using real-time PCR. *Environmental Science and Technology*, 43(3), 812-818.

Hawkins P. R., Chandrasena N. R., Jones G. J., Humpage A. R., Falconer I. A. (1997). Isolation and toxicity of *Cylindrospermopsis raciborskii* from an ornamental lake. *Toxicon,* 35(3), 314-346.

Herry, S. E., Fathalli, A., Rejeb, A. J., and Bouaicha, N. (2008). Seasonal occurrence and toxicity of *Microcystis* spp. And *Oscillatoria tenuis* in the Lebna Dam, Tunisia. *Water Research,* 42, 1263-1273.

Hotto, A. M., Satchwell, M. F., Berry, D. L., Gobler, C. J., and Boyer, G.L. (2008). Spatial and temporal diversity of microcystins and microcystin-producing genotypes in Oneida Lake, NY. *Harmful Algae*, 7, 671-681.

Humbert J. F., Quiblier C., Gugger M. (2010). Molecular approaches for monitoring potentially toxic marine and freshwater phytoplankton species. *Analytical and Bioanalytical Chemistry,* 397(5), 1723-1732.

Humpage, A. (2008). Toxin types, toxicokinetics and toxicodynamics. *Advances in Experimental Medicine and Biology*, 619, 383-415.

Joung, S., Oh, H., Ko, S., and Ahn, C. (2011). Correlations between environmental factors and toxic and non-toxic Microcystis dynamics during bloom in Daechung Reservoir, Korea. *Harmful Algae*, 10, 188-193.

Kaebernick, M., Dittmann, E., Börner, T., and Neilan, B. A. (2002). Multiple alternate transcripts direct the biosynthesis of microcystin, a cyanobacterial nonribosomal peptide. *Applied and Environmental Microbiology*, 68, 449-455.

Kim, S., Joung, S., Ahn, C., Ko, S., Boo, S. M., and Oh, H. (2010). Annual variation of *Microcystis* genotypes and their potential toxicity in water and sediment from a eutrophic reservoir. FEMS *microbiology*, 74, 93-102.

Kohl, J. G., Dudel, G., and Schlangstedt, M. (1985). On the morphological and ecological distriction of *Aphanizomenon flos-aquae Ralfs* ex Born et Flah and *Aphanizomenon graile (Lemm) Lemm. Archiv für Protistenkunde*, 130, 119-131.

Kumar, A., Kumar, A., Rai, A. K., and Tyagi, M. B. (2011). PCR-based detection of *mcy* genes in blooms of *Microcystis* and extracellular DNA of pond water. *African Journal of Microbiology Research*, 5, 374-381.

Kurmayer R., Kutzenberger T. (2003). Application of real-time PCR for quantification of microcystin genotypes in a population of the toxic cyanobacterium *Microcystis* sp. *Applied and Environmental Microbiology*, 69(11), 6723-6730.

Kurmayer, R., and Christiansen, G. (2009). The genetic basis of toxin production in cyanobacteria. *Freshwater Reviews*, 2, 31-50.

Kurmayer, R., Christiansen, G., Fastner, J., and Boner, T. (2004). Abundance of active and inactive microcystin genotypes in populations of the toxic cyanobacterium *Planktothix* spp. *Environmental Microbiology*, 6, 831-841.

Kurmayer, R., Dittmann, E., Fastner, J., and Chorus, I. (2002). Diversity of microcystin genes within a population of the toxic cyanobacterium *Microcystis* spp. in Lake Wannsee (Berlin, Germany). *Microbial Ecology*, 43, 107-118.

Larionov A., Krause A., Miller W. (2005). A standard curve based method for relative real time PCR data processing. *BMC Bioinformatics*, 6, 62.

Li, D., Kong, F, Shi, X., Ye, L., Yu, Y., and Yang, Z. (2012). Quantification of microcystin-producing and non-microcystin producing *Microcystis* populations during the 2009 and 2010 blooms in Lake Taihu using quantitative real-time PCR. *Journal of Environmental Sciences*, 24, 284-290.

Lindner, M., Bahnemann D. W., Hirthe, B., Griebler, W. D. (1995). Solar water detoxification: novel TiO2 powders as highly active photocatalysts. In: Stine, W. B., Tanaka, T., and Claridge, D. E. (Ed.), *Solar engineering* (pp. 399-407). New York: The American Society of Mechanical Engineers.

MacKintosh, C., and Klumpp, S. (1990). Tautomycin from the bacterium Streptomyces verticitlatus: another potent and specific inhibitor of protein phosphatases 1 and 2A. *FEBS Letters*, 277, 137-140.

McAlice, B. J. (1971). Phytoplankton sampling with the Sedgewick Rafter Cell. *Limnology and Oceanography*, 16, 19-28.

Mihali, T. K., Kellmann, R., Muenchhoff, J., Barrow, K. D., and Neilan, B. A. (2008). Characterization of the gene cluster responsible for cylindrospermopsin biosynthesis. *Applied and Environmental Microbiology*, 74, 716-722.

Mikalsen, B., Boison, G., Skulberg, O. M., Fastner, J., Davies, W., Gabrielsen, T. M., Rudi, K., and Jakobsen, K. S. (2003). Natural variation in the microcystin synthetase operon *mcy*ABC and impact on microcystin production in *Microcystis* strains. *Journal of Bacteriology*, 185, 2774-2785.

Moreira, C., Martins, A., Azevedo, J., Freitas, M., Regueiras, A., Vale, M., Antunes, A., and Vasconcelos, V. (2011). Application of real-time PCR in the assessment of the toxic cyanobacterium *Cylindrospermopsis raciborskii* abundance and toxicological potential. *Applied Microbiology and Biotechnology*, 92, 189-197.

Ngwa, F., Madramootoo, C., and Jabaji, S. (2012). Monitoring toxigenic *Microcystis* strains in the Missisquoi bay, Quebec, by PCR targeting multiple toxic gene loci. *Environmental Toxicology*.

Nishizawa, T., Asayama, M., Fujii, K., Harada, K., and Shirai, M. (1999). Genetic analysis of the peptide synthetase genes for a cyclic heptapeptide microcystin in *Microcystis* spp. *Journal of Biochemistry*, 126, 520-529.

Nubel U., Garcia-Pichel F., Muyzer G. (1997). PCR primers to amplify 16S rRNA genes from cyanobacteria. *Applied and Environmental Microbiology*, 63(8), 3327-3332.

Oh K. H., Jeong D. H., Shin S. H., Cho Y. C. (2012). Simultaneous Quantification of Cyanobacteria and *Microcystis* spp. Using Real-Time PCR. *Journal of Microbiology and Biotechnology*, 22(2), 248–255.

Ohtani I., Moore R. E., Runnegar M. T. C. (1992). Cylindrospermopsin: a potent hepatotoxin from the blue-green alga *Cylindrospermopsis raciborskii*. *Journal of American Chemical Society,* 114(20), 7941-7942.

Orr, P. T., and Jones, G. J. (1998). Relationship between microcystin production and cell division rates in nitrogen-limited *Microcystis aeruginosa* cultures. *Limnology and Oceanography*, 43, 1604-1614.

Orr, P. T., Rasmussen, J. P., Burford, M. A., Eaglesham, G. K., and Lennox, S. M. (2010). Evaluation of quantitative real-time PCR to characterize spatial and temporal variations in cyanobacteria, *Cylindrospermopsis raciborskii* (Woloszynska) Seenaya et Subba Raju and cylindrospermopsin concentrations in three subtropical Australian reservoirs. *Harmful Algae*, 9, 243-254.

Otsuka S., Suda S., Li R., Watanabe M., Oyaizu H., Matsumoto S., Watanabe M. M. (1998). 16S rDNA sequences and phylogenetic analyses of *Microcystis* strains with and without phycoerythrin. *FEMS Microbiology Letters,* 164(1), 119-124.

Otsuka, S., Suda, S., Watanabe, M., Oyaizu, H., Matsumoto, S., and Watanabe, M. M. (1999). Phylogenetic relationships between toxic and non-toxic strains of the genus *Microcystis* based on 16S to 23S internal transcribed spacer sequence. *FEMS Microbiology Letters*, 172, 15-21.

Ouahid, Y., and del Campo, F. F. (2009). Typing of toxinogenic *Microcystis* from environmental samples by multiplex PCR. *Applied Microbiology and Biotechnology*, 85, 405-412.

Ouahid, Y., Perez-Silva, G., del Campo, F. F. (2005). Identification of potentially toxic environmental *Microcystis* by individual and multiple PCR amplification of specific microcystin synthetase gene regions. *Environmental Toxicology*, 20, 235-242.

Paerl H. W., Huisman J. (2009). Climate change: a catalyst for global expansion of harmful cyanobacterial blooms. *Environmental Microbiology Reports*, 1(1), 27–37.

Pearson L. A., Neilan B. A. (2008). The molecular genetics of cyanobacterial toxicity as a basis for monitoring water quality and public health risk. *Current Opinion in Biotechnology*, 19(3), 281–288.

Philip O. T., Rasmussen J. P., Michele B. A., Geoffrey E. K., Sarah L. M. (2010). Evaluation of quantitative real-time PCR to characterise spatial and temporal variations in cyanobacteria, *Cylindrospermopsis raciborskii* (Woloszynska) Seenaya et Subba Raju and cylindrospermopsin concentrations in three subtropical Australian reservoirs. *Harmful Algae*, 9(3), 243-254.

Pidasak J. (1997). *Cylindrospermopsis raciborskii* (Woloszynska) Seenayya et Subba Raju, an expanding highly adaptive cyanobacterium: worldwide distribution and review of its ecology. *Archiv für Hydrobiologie Supplement*, 107, 563-593.

Rapala, J., Sivonen, K., Luukkanien, R., and Niemela, S. (1993). Anatoxin-a concentration in Anabaena and Aphanizomenon under different environmental conditions and comparison of growth by toxic and non-toxic Anabaena strains - a laboratory study. *Journal of Applied Phycology*, 5, 581-591.

Rasmussen, J. P., Giglio, S., Monis, P. T., Campbell, R. J., and Saint, C. P. (2008). Development and field testing of a real-time PCR assay for cylindrospermopsin-producing cyanobacteria. *Journal of Applied Microbiology*, 104, 1503-1515.

Rinta-Kanto, J. M., Konopko, E. A., DeBruyn, J. M., Bourbonniere, R. A., Boyer, G. L., and Wilhelm, S. W. (2009). Lake Erie *Microcystis*: Relationship between microcystin production, dynamics of genotypes and environmental parameters in a large lake. *Harmful Algae*, 8, 665-673.

Rinta-Kanto, J. M., Ouellette, A. J., Boyer, G. L., Twiss, M. R., Bridgeman, T. B., and Wilhelm, S. W. (2005). Quantification of toxic *Microcystis* spp. During the 2003 and 2004 blooms in western Lake Erie using quantitative real-time PCR. *Environmental Science and Technology*, 39, 4198-4205.

Rouhiainen, L., Vakkilainen, T., Siemer, B. L., Buikema, W., Haselkorn, R., and Sivonen, K. (2004). Genes coding for hepatotoxic heptapeptides (microcystins) in the cyanobacterium Anabaena strain 90. *Applied and Environmental Microbiology*, 70, 686-692.

Saker, M. L., and Neilan, B. A. (2001). Varied diazotrophies, morphologies, and toxicities of genetically similar isolates of *Cylindrospermopsis raciborskii* (nostocales, cyanophyceae) from Northern Australia. *Applied and Environmental Microbiology*, 67, 1839-1845.

Schindler, D. W. (1977). Evolution of phosphorus limitation in lakes. *Science*, 195, 260-262.

Scholin C. A., Herzog M., Sogin M., Anderson D. M. (1994). Identification of group- and strain-specific genetic markers for globally distributed Alexandrium (Dinophyceae) species. II. Sequence analysis of a fragment of the large-subunit ribosomal RNA gene. *Journal of Phycology*, 30, 999–1011.

Schoske R., Vallone P. M., Ruitberg C. M., Butler J. M. (2003). Multiplex PCR design strategy used for the simultaneous amplification of 10 Y chromosome short tandem repeat (STR) loci. *Analytical and Bioanalytical Chemistry*, 375(3), 333-343.

Senogles, P. J., Scott, J. A., Shaw, G., and Stratton, H. (2001). Photocatalytic degradation of the cyanotoxin cylindrospermopsin, using titanium dioxide and UV irradiation. *Water Research*, 35, 1245-1255.

Sivonen, K. (2008). cyanobacterial harmful algal blooms: state of the science and research needs. In: H. K. Hudnell (Ed.), *Emerging high throughput analyses of cyanobacterial toxins and toxic cyanobacteria*, (pp. 539-557). New York: Springer.

Sivonen K., Jones G. (1999). cyanobacterial toxins, in: Bartram ICaJ (ed) Toxic cyanobacteria in water: a guide to their public health consequences, monitoring, management (pp. 55-71). London: E and FN Spoon.

Sivonen, K., and Jones, G. (1999). cyanobacterial toxins. In: Chorus, I., Bartram, J. (Ed.), Toxic *Cyanobacteria in Water: A Guide to Their Public Health Consequences, Monitoring and Management* (pp. 41-111). London: E and FN Spon.

Song, L., Sano, T., Li, R., Watanabe, M. H., Liu, Y., and Kaya, K. (1998). Microcystin production of *Microcystis viridis* (cyanobacteria) under different culture conditions. *Phycological Research*, 46 (supplement), 19-23.

Srivastava, A., Choi, G., Ahn, C., Oh, H., Ravi, A. K., and Asthana, R. K. (2012). Dynamics of microcystin production and quantification of potentially toxigenic *Microcystis* sp. using real-time PCR. *Water Research*, 46, 817-827.

Te, S. H., and Gin, K. Y. (2011). The dynamics of cyanobacteria and microcystin production in a tropical reservoir of Singapore. *Harmful Algae*, 10, 319-329.

Tillett, D., Dittmann, E., Erhard, M., von Döhren, H., Börner, T., and Neilan, B. A. (2000). Structural organization of microcystin biosynthesis in *Microcystis aeruginosa* PCC7806: an integrated peptide-polyketide synthetase system. *Chemistry and Biology*, 7, 753-764.

Tillett, D., Parker, D. L., and Neilan, B. A. (2001). Detection of toxigenicity by a probe for the microcystin synthetae a gene (*mcy*A) of the cyanobacterial genus *Microcystis*: comparison of toxicities with 16S rRNA and phylogenies. *Applied Environmental Microbiology*, 67, 2810-2818.

Tomioka N., Nagai T., Kawasaki T., Imai A., Matsushige K., Kohata K. (2008). Quantification of *Microcystis* in a Eutrophic Lake by simple DNA extraction and SYBR Green real-time PCR. *Microbes and Environments*, 23(4), 306-312.

Vaitomaa, J., Rantala, A., Halinen, K., Rouhiainen, L., Tallberg, P., Mokelke, L., and Sivonen, K. (2003). Quantitative real-time PCR for determination of microcystin synthetase E copy numbers for *Microcystis* and *Anabaena* in lakes. *Applied and Environmental Microbiology*, 69, 7289-7297.

Valério, E., Chambel, L., Paulino, S., Faria, N., Pereira, P., and Tenreiro, R. (2010). Multiplex PCR for detection of microcystin-producing cyanobacteria from freshwater samples. *Environmental Toxicology*, 25, 251-260.

Welker, M., Burnke, M., Preussel, K., Lippert, I., and von Dohren, H. (2004). Diversity and distribution of *Microcystis* (Cyanobacteria) oligopeptide chemotypes from natural communities studied by single-colony mass spectrometry. *Microbiology*, 150, 1785-1796.

Westrick, J. A. (2008). Chapter 13: cyanobacterial toxin removal in drinking water treatment processes and recreational waters. In State of the Science and Research Needs. H. K.

Hudnell (Ed.), pp. 950, http://www.epa.gov/cyano_habs_symposium/monograph/Ch36.pdf

WHO (1998). Guidelines for drinking-water quality. Second edition, Addendum to vol 2, Health criteria and other supporting information. World Health Organisation, Geneva.

Wilson K. M., Schembri M. A., Baker P. D., Saint C. P. (2000). Molecular characterization of the toxic cyanobacterium *Cylindrospermopsis raciborskii* and design of a species-specific PCR. *Applied and Environmental Microbiology*, 66(1), 332-338.

Wittwer C. T., Herrmann M. G., Moss A. A., Rasmussen R. P. (1997). Continuous fluorescence monitoring of rapid cycle DNA amplification. *BioTechniques*, 22(1), 130-138.

Wiedner, C., Visser, P. M., Fastner, J., Codd, G. A., Mur, L. R., and Metcalf, J. (2003). Effects of light on the microcystin content of *Microcystis* Strain PCC 7806. *Applied and Environmental Microbiology*, 69, 1475-1481.

Zhang, W., Lou, I., Ung, W. K., Kong, Y., and Mok, K. M. (In press), Spatio-temporal variations of phytoplankton structure and water quality in the eutrophic freshwater reservoir of Macau. *Desalination and Water Treatment.*

Zhang W., Lou I., Ung W. K., Kong Y., Mok K. M. (2013). Eutrophication Analysis and Principle Component Regression for Two Subtropical Storage Reservoirs in Macau, *Desalination and Water Treatment*, 51(37-39): 7331-7340.

Zhang, W., Lou, I., Ung, W. K., Kong, Y., and Mok, K. M. (2014). Development of PCR and Real-time qPCR for Identifying and Quantifying Cyanobacteria, Microcystis *spp.* and C. *raciborskii* in Macau Storage Reservoir, *Frontiers of Earth Science*, 8(2): 291-301.

Biographical Sketch

Please fill out the Biographical Sketch listed here to have your abstract and biographical sketch published in Nova's series of books containing specialized-topic research summaries (abstracts) and biographical sketches:

Name: Inchio Lou

Affiliation: University of Macau

Education: Ph.D., North Carolina State University (NCSU), Environmental Engineering, May, 2006

M.S., Tsinghua University, P.R. China, Environmental Engineering, June, 2000

B.S., University of Macau, Macau SAR, Environmental Engineering, June, 1996

Address: Department of Civil and Environmental Engineering, Faculty of Science and Technology, University of Macau, Av. Padre Tomás Pereira Taipa, Macau SAR, China

Research and Professional Experience

- Microbial Ecology
- Fundamental Water/Wastewater Treatment Problems: Algal Bloom; Filamentous Bulking
- Environmental Bioprocess Technology

- Modeling in Water/Waster Treatment System; Water Quality Model
- Computational Toxicology and Risk Assessment

Professional Appointments

08/2011 – present	American Water Works Association (AWWA)
10/2010 – present	International Water Association (IWA)
11/2001 – present	Water Environment Federation (WEF)

Honors

2010, U.S. EPA Level III Scientific and Technological Achievement Award (#S10HE0050) for Innovative Modeling of Perfluorooctanoic Acid to Bridge Animals and Humans for Study Design and Risk Assessment.

Publications Last 3 Years

- Lou, Z. Xie, W. K.Ung, and K. M. Mok (in press), Freshwater Algal Bloom Prediction by Extreme Learning Machine in Macau Storage Reservoirs, *Neural Computing and Applications.*
- Y. Kong, I. Lou, Y. Zhang, C. U. Lou, K. M. Mok (in press), Using an Online Phycocyanin Fluorescence Probe for Rapid Monitoring of Cyanobacteria in Macau Freshwater Reservoir, *Hydrobiologia.*
- W. Zhang, I. Lou, W. K. Ung, Y. Kong and K. M. Mok (in press), Analysis of Cylindrospermopsin-and Microcystin- Producing Genotypes and Cyanotoxin Concentrations in the Macau Storage Reservoir, *Hydrobiologia.*
- W. Zhang, I. Lou, W. K. Ung, Y. Kong and K. M. Mok (in press), Application of PCR and Real-time PCR for Monitoring Cyanobacteria, *Microcystis* spp. and *Cylindrospermopsis raciborskii* in Macau Freshwater Reservoir, *Frontiers of Earth Science.*
- Z. Li, I. Lou, H. Lei and Y. Liu (in press), Effects of alumina trihydrate (ATH) on formation of biofilms settled on inert carriers of polyethylene terephthalate (PET), *Desalination and Water Treatment.*
- Y. Yu., I. Lou, Y. X. Li (2014), The current situation and prospection and the prospect of water resource in Macau, *Water and Wastewater Engineering* (Chinese core journal organized by by the Ministry of Construction of construction of China), 40 (2): 30-33.
- W. Zhang, I. Lou, W. K. Ung, Y. Kong and K. M. Mok (2013), Eutrophication Analysis and Principle Component Regression for Two Subtropical Storage Reservoirs in Macau, *Desalination and Water Treatment*, 51(37-39): 7331-7340.

- Lou, Y. Zhao, Y. Wu and P. F. Ricci (2013), Systems cancer biology and the controlling mechanisms for the J-shaped cancer dose response: towards relaxing the LNT hypothesis, *Dose-Response*, 11 (3): 301-318.
- Lou, S. Gong, X. Huang, Y. Liu and K. M. Mok (2013), Coagulation optimization using ferric and aluminum salts for treating high algae and high alkalinity source water in a typical North-China plant, *Desalination and Water Treatment*, 51(16-18): 3361-3370.
- Lou and Y. Zhao (2012), Sludge bulking prediction using principle component regression and artificial neural network, Mathematical Problems in Engineering, 2012, Article ID 237693, 17 pages.
- Z. Xie, I. Lou, W. K. Ung and K. M. Mok (2012), Freshwater Algal Bloom Prediction by Support Vector Machine in Macau Storage Reservoirs, Mathematical Problems in Engineering, 2012, Article ID 397473, 12 pages.
- Lou, S. Gong, X. Huang, and Y. Liu (2012), Coagulation optimization for low temperature and low turbidity source water using combined coagulants: A case study, *Desalination and Water Treatment,* 46(1-3): 107-114.
- Y. Zhao, I. Lou, and R. B. Conolly, (2012), Computational modeling of signaling pathways mediating cell cycle and apoptotic responses to ionizing radiation-induced DNA damage, *Dose-Response*, 10(2): 251-273.
- C. N. Lei, I. Lou, H. U. Song and P. Sun (2012), Turbidity removal improvement for Yangze River raw water, *Desalination and Water Treatment*, 45(1-3): 215-221.
- Z. Li, H. Lei, and I. Lou (2012), Combined effects of flow rate and light on characteristics of biofilms grown on three-dimensional elastic carriers, *Desalination and Water Treatment*, 45(1-3): 241-249.

In: Cyanobacteria
Editor: Douglas Davison

ISBN: 978-1-61470-925-1

Chapter 5

REVIEW: ADVANCES IN MICROBIOLOGICAL DEGRADATION OF MICROCYSTINS

Melina Celeste Crettaz-Minaglia, Darío Andrinolo and Leda Giannuzzi[*]
Toxicology Laboratory, Exact Sciences College,
National University of La Plata (UNLP), La Plata, Argentina

ABSTRACT

Microcystins (MCs) are secondary metabolites produced by freshwater cyanobacteria that have been associated with severe episodes of human and animal acute hepatotoxicity. MC is produced by several cyanobacteria species from t different genera such as *Microcystis, Anabaena, Planktothrix.* These have worldwide distribution in freshwater bodies and are known for their hepatotoxic effects in mammals by inhibiting protein phosphatase. So far, concern is growing over MCs contamination in drinking water in many countries around the world. MC is recalcitrant to conventional water treatment and is often detected even after chemical treatment of waters. Chemicals used to kill the algae are potentially toxic and may remain in the water to pose greater danger when consumed. Consequently, the biological methods of removing MCs are an alternative to the disadvantages of using chemicals.

In this chapter, a review of the literature on the bacterial degradation of MCs was performed. Methods of isolation of bacteria were collected and a review of the state of knowledge of the subject was made including degradation mechanisms. Furthermore, we have exposed the advantages and disadvantages of biological methods and their potential application to water treatment biotechnology.

Highlights: Bacterial degradation of microcystins and their potential application to water treatment technology

Keywords: Bacteria degradation, Microcystins, biological methods, water treatment

[*] Email: leda@biol.unlp.edu.ar.

INTRODUCTION

The microcystins (MCs) are toxic cyclic heptapeptides and contain an unusual B-amino acid (3 - amino - 9 - methoxy - 10 - phenyl - 2,3,8 - trimethyl - deca-4,6-dienoic acid) (Adda). They are produced by some genera of cyanobacteria from freshwater such as *Microcystis, Planktothrix, Anabaena, Nostoc* and *Snowella* [1]. Although MC-LR is the most commonly found MC [2], many other MCs are known, and blue-green algae which produce hepatotoxins usually produce a mixture of different MCs [3]. These potent mammalian toxins are known to be responsible for acute and chronic effects which cause liver haemorrhage within few hours and show tumor promotion activity through protein phosphatase inhibition observed in rodents as well as in primary hepatocytes in vitro. The resulting hyperphosphorylation of intracellular protein leads to disruption of intermediate filaments that form the cellular scaffold in human and rodent hepatocytes [2].

The first confirmed case of human death caused by cyanotoxins occurred in 1996, over 100 patients developed acute liver failure and over 50 persons died from using water contaminated with MCs for hemodialysis in Caruaru, Pernambuco State (Brazil) [4,5]. Biological and chemical analyses confirmed the presence of MCs and cylindrospermopsin in the activated carbon filter used in the clinic's water purification system. MCs were also detected in samples of blood and liver tissue from the affected patients [4, 6, 7].

The International Agency for Research on Cancer (IARC) classifies MC-LR as possibly carcinogenic to humans (group 2B), based on the inadequate evidence that it directly cause cancer in either laboratory animals or humans [8]. Furthermore, Tian et al. [9] reported investigations that suggest the MCs may be responsible for the high incidence of liver cancer in populations dependent upon MC-contaminated drinking water in China [10], Serbia [11], and Florida of the United States [12], as well as colorectal cancer in China [13].

As a result of the reported effects of MCs, the World Health Organization (WHO) considered MC-LR as one of the most frequent cyanotoxins, and proposed a guideline of $1\mu g.L^{-1}$ as the maximum concentration in drinking water [14].

Likewise, it is known that MCs are toxins resistant to conventional process employed in the water treatment plants [15] and often detected after traditional treatments such as chemical or physical process. Furthermore, these may require a more costly treatment such as activated carbon and/or advanced oxidation processes, which themselves may have limitations [16]. Poor or no MCs removal by coagulation, flocculation- sedimentation, filtration and chlorination has been reported [17]. These treatments are effective for removing intracellular toxins with intact cells without causing additional release of intracellular toxins [18, 19], however, they are inefficient at removing dissolved toxins [20, 21].

The presence of MCs in drinking water supply was detected in 10 of 13 samples in the cities of Ensenada and La Plata, Argentina [22]. Between December 2004 and April 2005, the maximum level was $6.7\ \mu g.L^{-1}$, this shows that the conventional water treatment process was inefficient, at least during the critical periods of massive development of cyanobacterial blooms.

About the use of chemical oxidants, the removal of toxins by chlorine was found to be pH-dependent. The MCs destruction decreased markedly at pH above 8. At room temperature water, the chlorine doses required to maintain a residual in the distribution system, it may not be enough to remove MCs and thus result in free toxin in the drinking water. A pre-treatment with oxidants (chloride, ozone, potassium permanganate) added prior to conventional treatment can alter the surface characteristics and charge of cyanobacteria which result in enhancing the effectiveness of coagulation and flocculation for their subsequent removal [23, 24], [25, 26], and [27, 28]. However, the oxidants also cause cell lysis, and effective control on particular concentrations are required for effective pre-oxidation. For example, chlorine pre-oxidation could induce cell lysis and concomitant toxin release [29]. Another disadvantage of pre-chlorination is the formation of chlorinated organic molecules when the organic material is present. These products described as disinfection by-products include a range of chloride compounds toxic to mammals including carcinogenic trihalomethanes (THMs) recognized as harmful [21].

Similarly, ozonation can lead to bromate formation which is a potential carcinogen [30]. Natural organic material can consume ozone thereby reducing its concentration in solution [31]. Moreover, the cost of using ozone is relatively high and it has inherent complexity in operation in water treatment plants.

In contrast, little is known about the possibility of hazardous by-products caused by $KMnO_4$ whilst overdosing must be avoided since a residual $KMnO_4$ of 0.05 mg. L^{-1} or greater will result in a pink taint on drinking water [32].

Other method for removing MCs in water treatment plants is activated carbon. Granular activated carbon was shown to be effective in toxins removal, and the efficiency depends on the type of activated carbon and the water quality conditions. However, carbon became progressively saturated, and less adsorbent action was observed as their volume filtered increased [33].

The presence of natural organic material reduces the effectiveness of these treatment processes for the removal of MCs. Combination of adsorption of organic material, and cyanotoxins on the surface of the carbon has been observed in pilot studies, and there was a competition for adsorption sites of carbon [34]. Recently, Zhang et al. [35] investigated a new kind of low-cost syntactic adsorbent from bamboo charcoal and chitosan was developed for the removal of MC-LR from drinking water. However, it presents the same disadvantages as the absorbent traditional, but has a lower cost.

In other way, biological processes have been used for centuries in treatment of drinking water and, in general, they are low technology processes that generally require little or no maintenance or running costs, and have the advantage of being a "natural" treatment [16].

Currently, biological methods of MCs removal are being widely investigated [3, 36], in particular, biological filtration processes, for cyanobacteria metabolite removal [16]. Many authors have reported several MCs degrading bacteria [37, 38, 39, 40]. Ishii et al. [41] proposed that the biodegradation is one of the safest and mildest treatments for removing cyanobacterial toxins from water. This is, partly due to the fact that there is no evidence of MCs accumulation in the environment and its stability in relation to a wide range of physicochemical conditions, which suggests that biodegradation of these toxins, is the major route for its elimination from the environment [36, 42].

This review focuses on the MCs bacterial degradation as part of biological methods and their technological potential to be applied in the drinking water treatment.

CURRENT STATE OF KNOWLEDGE

The MCs biodegradation has been studied in drinking water sources, sediments and active biofilms. There are few studies in natural environments perhaps because there is no technological application.

On one hand, it has been studied MC-LR microbial degradation by indigenous mixed bacterial populations [3]. These authors identified the degradation products and suggest that the Adda side chain in MC-LR is the site of the initial biodegradative attack.

Furthermore, there is a growing number of isolated organisms reported as having the ability to degrade MCs in water and so far the majority appears to belong to the family *Sphingomonadaceae* [16] The majority of these MC-degrading organisms have been identified as genera *Sphingomonas* [43, 44, 45, 46, 47, 41, 48, 37] and *Sphingopyxis* [49, 42, 50].

Almost all of the genotypic studies on MCs degradation have focused on the *Sphingomonadaceae* and have shown that they contain specific genes required for MCs degradation.

Within the genome of the first isolated MC-degrading bacterium, *Sphingomonas* sp. ACM-3962, Bourne et al. [43, 44] identified a gene cluster, *mlrA, mlrB, mlrC* and *mlrD*, involved in MC-LR degradation. The authors determined that the *mlrA* gene encoded an enzyme responsible for the hydrolytic cleaving of the cyclic structure of MC-LR. The initial hydrolysis results in a substantial reduction of molecular toxicity [51]. The mlrA gene sequence is very rare, without homologues found in the public databases, and therefore its functional characteristics are difficult to be assigned [51]. The resultant linear MC-LR molecule was then sequentially hydrolysed by peptidases encoded by the *mlrB* and *mlrC* genes. The final gene, *mlrD*, encoded for a putative transporter protein that may have allowed MCs active transport and/or its degradation products into or out of the cell. The ability to degrade MCs is not commonly present in the *Sphingomonas* genus, but only in specific bacterial strain [46].

The studies have demonstrated that MCs biodegradation do not yield toxic by-products. Thereby Bourne et al. [43] and Harada et al. [47] identified two intermediate products from the MC-LR bacterial degradation by *Sphingomonas* sp. ACM-3962 and *Sphingomonas* sp. B9, respectively. Both studies identified linearised MC-LR (NH2-Adda-Glu-Mdha-Ala-Leu-MeAsp-Arg-OH) and a tetrapeptide (NH2-Adda-Glu-Mdha-Ala-OH) as the intermediate products, and Harada et al. [47] also isolated Adda as one of the final degradation products.

Amé et al. [37] studied MC-RR biodegradation pathway by *Sphingomonas* sp. CBA4 strain would start by demethylating MC-RR, affording an intermediate product. MC-RR degradation was also investigated by Wang et al. [50]. Two intermediate metabolites and a dead-end product of MC-RR were observed on HPLC profiles and all of them had similar scanning profiles in the wavelength from 200 to 300 nm, indicating that the group of Adda in all products of MC-RR remained intact. Yan et al. [52] characterized the first step involved in enzymatic pathway for MC-RR biodegraded by *Sphingopyxis* sp. USTB-05. The nucleotide sequences of cloned USTB-05-A possessed 92.5% homology to that of *mlrA* reported in *Sphingomonas* sp. strain ACM-3962. The deduced amino acid sequences containing the cleavage sites of 26th (alanine) and 27th (leucine) showed 83% identical to that of *mlrA*. Yan et al. [53] cloned and expressed the first gene for biodegrading MC-LR by *Sphingopyxis* sp. USTB-05. This bacterium was shown to biodegrade MC-LR [50].

Maruyama et al. [54] described *Sphingosinicella microcystinivorans* gen. nov. sp. nov. (*Sphingomonadacea*) strain Y2^T, MDB2, and MDB3. They were isolated from a eutrophic lake (Japon).

Moreover, *Pseudomonas* sp. has been reported as a bacterium able to degrade MC-LR [55, 56]. This toxin was metabolized to (2S, 3S, 8S)-3- amino-2, 6, 8-trimethyl-10-phenyldeca-4E, 6E-dienoic acid (DmADDA) and concluded that MC was degraded mainly by the action of *P. aeruginosa* alkaline protease. Kang et al. [57] studied 30 strains of *Pseudomonas*. *Pseudomonas aeruginosa* PA14 exhibited higher degradation activity. These authors determined that the activity of PA14 diminished when its density decreased to less than 10^6 cells.mL^{-1}. Rapala et al. [58] reported a novel genus and species; *Paucibacter toxinivorans* capable of degrading MCs. Hu et al. [59] isolated a *Methylobacillus* sp. strain J10 from a cyanobacterial sludge which could effectively degrade MC -LR and MC-RR. Eleuterio and Batista [60] identified a bacterium capable of degrading MC-LR, *Morganella Morganii*. This bacterium was found in both the anthracite media from the Los Angeles filtration plant and in Lake Mead water. Analysis by qPCR confirmed the presence of the *mlrA* gene. Chen et al. [61] isolated a bacterium *Stenotrophomonas* sp. strain EMS with the capability of degrading MC-LR and –RR. Ramani et al. [62] also isolated *Rhizobium gallicum* (DC7) capable of degrading MC-LR.

On the other hand, Manage et al. [63] reported the first evidence of MCs degraders that do not belong to the *Proteobacteria* identified as *Arthrobacter* sp., *Brevibacterium* sp., and *Rhodococcus* sp. However, the *mlr* gene cluster was not detected [64]. Furthermore, they have also been studied Actibacteria. In these studies they have demonstrated that probiotic bacteria have the ability to biologically remove MCs [65, 66]. Ramani et al. [62] isolated *Microbacterium* sp (DC8) capable of degrading MC-LR from Lake Okeechobee. The probiotic bacteria, *Lactobacillus rhamnosus* strains GG and LC-705, *Bifidobacterium longum* 46, *Bifidobacterium lactis* strains 420 and Bb12, were all able to remove MC-LR, MC-RR, and MC-LF [65, 66]. However, they were unable to completely remove MCs. The MC-LR maximum removal, 60.3%, was observed with *L. rhamnosus* GG, of MC-RR, 62.8%, and MC-LF, 77.4% with *L. rhamnosus* LC-705 (toxin concentration of 100 mg.L^{-1}, at 37°C, in 24 hours).

Meanwhile, Nybom et al. [67] confirmed the hypothesis of MC-LR enzymatic degradation in presence of probiotic lactic acid bacterial and bifidobacterial strains, and the participation of the proteolytic system of the bacteria in this process. The results suggest that extracellular proteinases are involved in the MC-LR decomposition.

Hu et al. [68] isolated a bacterium EMB identified as *Bacillus* sp. In this strain, *mlrA*, MC-degrading gene was detected. Alamri [69] isolated a bacterium identified as *Bacillus flexus* strain SSZ01 and *mlrA* gene was detected. In addition, this strain could be MC-RR as carbon and/ or nitrogen source.

It has been reported a link between previous exposure to MCs and the rate of degradation [70, 71]. Giaramida et al. [72] suggested that MCs exposure drives changes in structure and physiology of bacterial communities and in turn those communities differentially perform MC-LR degradation. Moreover, dissolved organic carbon, nitrogen compounds, and temperature had a significant effect in shaping bacterial communities structure [72]. Chen et al. [73] informed that sediments play a crucial role as a source (to water column) of biodegrading bacteria and as a carbon-rich environment for bacteria to proliferate and metabolize MCs and other biogenic toxins produced by cyanobacteria. They concluded that alkali-tolerant MCs degrading bacterium played a key role in triggering the rapid MCs degradation, leading to the disappearance of toxic water blooms in aquatic environments [49]. Park et al. [45] indicated that the degradation rates were strongly dependent to temperature and the maximum rate was at 30°C. The authors suggested that strain called Y2 (*Sphingomonadacea*) could use MC as a carbon and energy source.

Eleuterio and Batista [60] concluded that an additional carbon source, as biodegradable organic matter, significantly inhibits the MCs degradation.

Moreover, due to numerous non-culturable bacteria that exist in the natural environment, exceeding 99% of the total bacterial count in various systems [73], they have studied the degradation of MC by cultura-independent techniques [74]. The authors examined the process of MCs degradation by bacteria in the natural environment, focusing on the population dynamics of MC-degrading bacteria and used FISH method (fluorescence *in situ* hybridization) to analyze the population dynamics of MC-degrading bacteria in *Microcystis* mucilage. In this investigation, members of the *Cytophaga/Flavobacterium* group were able to degrade *Microcystis* cells.

According to Kormas and Lymperopoulou [64] the ubiquitous α- and β-*Proteobacteria* and *Actinobacteria* include the majority of potential cyanotoxin-degrading bacteria. However, their research shows that only few bacterial species bear the *mlrA* gene and these bacteria is not abundant in the ecosystems where MCs are found.

Bacteria Isolated Methods

Isolation methods of potentially MC-degrading bacteria have been collected from the literature available. These methods assume that the bacteria use the MCs as a carbon or nitrogen source.

Potentially MCs degrading bacteria have been isolated from surface water [56, 45, 74, 41, 38, 37, 39, 49, 63, 60, 62, 69], sediment (38, 59, 42, 61, 68] and active biofilm [76]. Usually the bacteria came from places that have a history of cyanobacterial blooms. The samples were taken in sterile bottles and stored at 4°C. These can be filtered by filters of different porosity: 40 μm [45, 60], 10 μm cellulose acetate filters [56], GF/C glass-fiber filters mean pore diameter 1.2 μm [49] or unfiltered.

Different types of media were used to growth bacteria, for example Powell & Errington´s medium [60], aspargine liquid culture medium [56], nutrient agar (NA medium, 1% agar plate) [45], M9 liquid medium [41], Sakurai medium [38, 76], mineral salts medium (MSM) with MC as the sole carbon and nitrogen source [37, 39, 42, 53, 62], Enrich Bacteria Broth (EBB) medium with MC extract [59, 61, 68], water samples with MC in 0.1M carbonate buffer (pH 9.5) and an antibacterial-antimycotic mixed stock solution (Nacalai Tesque Japan) containing final concentrations of penicillin (10 unit.mL^{-1}), streptomycin (100 μg.mL^{-1}), and amphotericin B (0.25 μg.mL^{-1}) [49]. In all the cases, details of the media are present in the bibliography section.

In general, the culture conditions should be similar to the natural environment where the samples were collected. Particularly it can be observed that the temperature is an important factor. The growth conditions can be with shaking (100 rpm [63], 120 rpm [68, 59, 69], 200 rpm [76, 50] or without shaking, in dark, at different temperatures as 23°C [37], 25°C [49, 63 77], 26 °C [62], 27°C [45, 38, 76], 26-28 °C [39], 28°C [69], 30°C [41, 59, 42, 61, 50, 68] and, 37°C [56].

The isolations were made in plates with different medium as Powell and Errington´s medium [60], acetamide agar slant medium [56], nutrient agar (NA medium, 1% agar plate) [45, 42], Sakurai medium [38], tryptone soy agar supplemented with MC-RR [38], peptone-yeast (PY) medium [49], Mineral Salts (MS) agar medium with 20% (v/v) of crude MCs extract [61], LB agar [63], Nutrient Broth Agar [69], solid R2A medium [77]. The colonial specimens grown on each original plate were streaked and several transfers were performed until single and pure colonies were obtained [41, 38, 63, 60]. The single colonies from these plates were transferred back to a liquid medium and they allowed to grow and obtain a sufficient number of bacteria for DNA extraction [60]. Takenaka and Watanabe [56] identified the single colony with API 20NE (Bio Mérieux S.A., France) and Amé et al. [37] and, Alamri [69] characterized considering 16S ribosomal RNA sequence analysis.

POTENTIAL APPLICATION

Cyanobacterial blooms are a great challenge for drinking water production, because of their occurrence in drinking water resource often causes several process disturbances in treatment plant, such as faster filter clogging, and reactant consumption increase [78], and more expensive processes. Consequently, the MCs effective removal from potable drinking water is a major goal for all the water utilities [79].

Water treatment costs combined with water scarcity and increasing water demand present a huge problem in the developing world where populations are frequently exposed to cyanobacterial toxins amongst other organic and microbial contaminants. Thus, there is need for simple, low cost and effective water treatment technology [80].

Although many methods, such as biomanipulation and cyanocides, have been tried for the cyanobacteria elimination in a lake, no suitable method has been developed, and it would be difficult to regulate the occurrence of cyanobacteria by conventional methods [81].

The bacterial strains or possible enzymes identified in the removal process could be used in a degradation process to remove toxins from drinking water. Chen et al. [61] mentioned that the key to bioremediate MCs-contaminated water is employ native bacteria in order to avoid biotic intrusion. However, as in any biological system, biodegradation is susceptible to many factors such as the temperature, the initial inoculum of bacteria, the presence of organic matter and the presence of compounds such as metals, but also the pre-exposure to the toxin that can eliminate or diminish the lag phase [16].

According to Ho et al. [16], biological filtration processes can be the best way to use organism for the cyanobacterial metabolites biodegradation. They also added that the operational conditions are important for the successful cyanobacterial metabolites removal. Ho et al. [48] did not found any difference between the retention times of 4 minutes and 30 minutes in the MCs biodegradation through sand filters.

This has the advantage that most of the treatment processes already employ a filtration step.

Other support used was granular activated carbon (GAC). The GAC filters offer the advantage of two removal mechanisms, adsorption and biodegradation, and thus are an attractive treatment option for effective removal for organic contaminants [16]. In addition to adsorption, toxins degradation can be performed by biofilm on the surface in the filter. MCs removal can be substantially better by biological degradation in addition to adsorption [82].

Tsuji et al. [38] investigated the MC-LR and MC–RR degradation in a cyanobacterium, NIES-102, using immobilized *Sphingomonas* B-9. MCs Degradation with immobilizing carriers such us cellulose, polyester or entrapped immobilization (Polyethylene glycol gel) on strain B-9 were studied. These experiments showed that polyester was the most suitable carrier for the adhesive immobilization of the B-9 strain. Immobilized B-9 showed over 80% removal MC continued for 2 months under the optimized conditions.

It has been reported in the literature the use of biofilms and sand filters that degrade MCs. These could be used routinely as a strategy of low-cost removal treatment systems for drinking water [44, 48, 79, 83].

Ho et al. [48] were able to provide evidence that the MCs removal through a sand filter was predominantly through biological processes. MCs complete removal was founded under rapid sand filtration conditions, and showed to be primarily through biological degradation rather than any physical processes. Detection of the bacterial *mlrA* genes in sand filters were reported by Ho et al. [79]. Thereon, the studies regarding treatment plants for drinking water have shown the importance of biofilm particularly in filters. Thus, Hoeger et al. [51] verified that the efficiency of two different water treatment systems with respect to the reduction of cyanobacterial toxins, the MCs concentrations in water samples from surface waters and their associated water treatment plants in Switzerland and Germany were investigated. Toxin concentrations in samples from drinking water treatment plants ranged from below 1.0 μg to more than 8.0 $\mu g.L^{-1}$ MC-LR equiv.L^{-1} in raw water and were distinctly below 1.0 $\mu g.L^{-1}$ after treatment.

Menwhile, Eleuterio and Batista [60] isolated MCs-degrating bacteria from an operating anthracite biofilter. The implication of this finding is that if biofilters can be used to remove MCs, there would be no need to bio-augment the filters with external bacterial sources because they are already present [60]. In this study, bacterial cultures that come from a biofilm, that were formed on anthracite surface, were able to degrade MCs aerobically using the toxin as the sole carbon source. The authors concluded that the feasibility of using biofilters to remove MCs from water largely depends on the biodegradable naturally organic matter and toxin concentration in the influent water.

Moreover, the fate of multiple cyanobacterial metabolites was assessed in two Australian source waters by Ho et al. [77]. The saxitoxins were the only metabolites shown to be non-biodegradable in Myponga Reservoir water, while MC-LR and geosmin were biodegradable in this water source. Likewise, cylindrospermopsin (CYN) was shown to be biodegradable in River Murray water. The facility order of biodegradability followed the trend: MC-LR > CYN > geosmin > saxitoxins. Biodegradation of the metabolites was affected by temperature and seasonal variations with more rapid degradation at 24 °C and during autumn compared with 14 °C and during winter. A MC-degrading bacterium (*Sphingopyxis* sp. strain TT25). This strain could degrade the MCs in the presence of copper sulphate (0.5 $mg.L^{-1}$ as Cu^{2+}) which is advantageous for water authorities dosing such cyanocides into water bodies to cyanobacterial blooms control. Meanwhile, Li et al. [84] studied the biofilm from water treatment plant near the lake Kasumigaura (Japan) that employs a biological treatment facility packed with vinyl chloride board as a carrier for biofilms. These authors confirmed that microcystin-LR (MCLR) was effectively biodegraded within 5.5 days in cultures of the biofilm sampled in winter, under controlled laboratory conditions. In conclusion, further research should focus on the development of new technology using efficient bacteria with higher activity to remove MC and improve immobilization techniques to be applied in treatment plant of drinking water.

This is an important aspect to consider in the future.

REFERENCES

[1] D. Andrinolo and D. Sedan. In: L. Giannuzzi, Ministerio de Salud, 41 (2011).

[2] W. W. Carmichael, *J. Appl. Bact.,* 72, 445 (1992).

[3] I. T. Cousins, D. J. Bealing, H. A. James and A. Sutton, *Water Research*, 30, 481 (1996).

[4] E. M. Jochimsen, W. W. Carmichael, J. An, M. C. Denise, S. T. Cookson, C. E. M. Holmes, M. B. C. F. D. A. Antunes, T. M. Lyra Melo, V.S.T. Barreto, S.M.F.O. Azevedo and W. R. Jarvis, *N. Engl. J. Med.,* 13, 873 (1998).

[5] S. Pouria, A. De Andrade, J. Barbosa, R. L. Cavalcanti, V. T. Barreto, C. J. Ward, W. Preiser, G. K. Poon, G. H. Neild and G. A. Codd, 352, 21 (1998).

[6] W.W. Carmichael, S.M.F.O. Azevedo, J.S. An, R.J.R. Molica, E.M. Jochimsen, S. Lau, K.L. Rinehart, G.R. Shaw, G.K. Eaglesham, *Environ. Health Perspect.,* 109, 663 (2001).

[7] S. M. F. O. Azevedo, W. W. Carmichael, E. M. Jochimsen, K. L. Rinehart, S. Lau, G. R. Shaw and G. K. Eaglesham, *Toxicology* 181-182, 441 (2002).

[8] IARC. Monographs volume 94. Available at http:// monographs.iarc.fr/ENG/ Meetings/94-cyanobacterial.pdf. (2006).

[9] D. Tian, W. Zheng, X. Wei, X. Sun, L. Liu, X. Chen, H. Zhang, Y. Zhou, H. Chen, H. Zhang, X. Wang, R. Zhang, S. Jiang, Y. Zheng,G. Yang and W. Qu, *Chemosphere,* 91, 1064 (2013).

[10] Y. Ueno, S. Nagata, T. Tsutsumi, A. Hasegawa, M. F. Watanabe, H. D. Park, G. C. Chen, G. Chen and S. Z. Yu, *Carcinogenesis,* 17, 1317 (1996).

[11] Z. Svircev, S. Krstic, M. Miladinov-Mikov, V. Baltic and M. Vidovic, *J. Environ Sci. Health* 27, 36 (2009).

[12] L. E. Fleming, C. Rivero, J. Burns, C. Williams, J. A. Bean, K. A. Shea and J. Stinn, *Harmful Algae*, 1, 157 (2002).

[13] L. Zhou, H. Yu and K. Chen, *Biomed. Environ. Sci.* 15, 166 (2002).

[14] World Health Organization, Guidelines for drinking-water quality. Vol. 2. Health criteria and other supporting information. Geneva. (1998).

[15] K. Lahiti and L. Hiisverta, *Water Supply.*, 7, 149 (1989).

[16] L. Ho, E. Sawade and G. Newcombe, *Water Research*, 46, 1536 (2012).

[17] A. M. Keijola, K. Himberg, A. L. Esala, K. Sivonen and L. Hiisvirta, *Toxicity Assessment*, 3, 643 (1988).

[18] A. K. Y. Lam, E. E. Prepas, D. Spink and S. E. Hrudey, *Water Research*, 29, 1845 (1995).

[19] C. W. K. Chow, M. Drikas, J. House, M. D. Burch and R. M. A. Velzeboer, *Water Research*, 33, 3253 (1999).
[20] K. Himberg, A. M. Keijola, L. Hiisvirta, H. Pyysalo and K. Sivonen, *Water Research*, 23, 979 (1989).
[21] C. Svrcek and D. W. Smith, *Journal of Environmental Engineering and Science*, 3, 155 (2004).
[22] R. Echenique, J. Rodríguez, M. Caneo, L. Giannuzzi, M. Barco, J. Rivera, J. Caixach and D. Andrinolo. Microcystins in the drinking water supply in the cities of Ensenada and La Plata (Argentina). In Aplicações da Ficologia: Anais do XI Congresso Brasileiro De Ficologia e Simpósio Latino-Americano sobre Algas Nocivas, 2006, Itajaí, SC. Rio de Janeiro: Museu Nacional: 141-148. (Série Livros, 30).
[23] J. D. Plummer and J. K. Edzwald, *Journal of Water Supply: Research and Technology e Aqua*, 51, 307 (2002).
[24] M. Ma, R. P. Liu, H. J. Liu, J. H. Qu and W. Jefferson, *Separation and Purification Technology*, 86, 19 (2012)
[25] H. F. Miao and W. Y. Tao, *Separation and Purification Technology*, 66, 187 (2009).
[26] J. J. Chen, H. H. Yeh and I. C. Tseng, *Journal of Chemosphere*, 74, 840 (2009).
[27] D. Ellis, C. Bouchard and G. Lantagne, *Desalination*, 130, 255 (2000).
[28] J. J. Chen and H. H. Yeh, *Journal of Water Science & Technology: Water Supply*, 6, 79 (2006).
[29] H. G. Peterson, S. E. Hrudey, I. A. Cantin, T. R. Perley and S. L. Kenefick, *Water Research*, 29, 1515 (1995).
[30] S. Bouland, J. P. Duguet and A. Montiel, Science and Engineering, 26, 381 (2004).
[31] L. Ho, Ph.D. Dissertation, University of South Australia, Adelaide, Australia (2004).
[32] W. A. Welch, Journal of the American Water Works Association, 55, 735 (1963).
[33] I. R. Falconer, M. T. C. Runnegar and T. Buckley, *J. Am. Water Works Assoc*, 18, 102 (1989).
[34] G. Jones, W. Minato, K. Craig and R. Naylor, Proceedings of the Australian Water and Wastewater Association 15th federal convention, (1993)
[35] H. Zhang, G. Zhu, X. Jia, Y. Ding, M. Zhang, Q. Gao, C. Hu and S. Xu, *Journal of Environmental Sciences,* 23 (12), 1983 (2011).
[36] C. Edwards and L. A. Lawton. In: A. I. Laskin, G. M. Gadd, S. Sariaslani (Eds). *Academic Press*, 109 (2009).
[37] V. Amé, E. J. Echenique, P. Stephan and W. D. Alberto, *Biodegradation* 17, 447 (2006).
[38] K. Tsuji, M. Asakawa, Y. Anzai, Y. Sumino and K. Harada, *Chemosphere*, 65, 117 (2006).
[39] G. A. F. Lemes, R. Kersanach, L. S. Pinto, O. A. Dellagostin, J. S. Yunes and A. Malthiensen, *Ecotoxicology and Environmental Safety*, 69, 358 (2008).
[40] M. M. Pathmalal C. Edwards, B. K. Singh and L. A. Lawton, *Appl Environ Microbiol.*, 75, 6924 (2009).
[41] H. Ishii, M. Nishijima and T. Abe, *Water Research*, 38, 2667 (2004)
[42] M. Zhang, G. Pan and H. Yan, *Journal Environmental Scienses,* 22, 168 (2010).
[43] D. G. Bourne, G. Jones, R. Blakeley, A. Jones, A. Negri and P. Riddles, *Appl. Environ. Microbiol.*, 62, 4086 (1996).

[44] D. G. Bourne, P. Riddles, G. J. Jones, W. Smith and R. L. Blakeley, *Environ Toxicol.,* 16, 523 (2001).
[45] H. D. Park, Y. Sasaki, T. Maruyama, E. Yanagisawa, A. Hiraishi and K. Kato, *Environ Toxicol*, 16, 337 (2001).
[46] T. Saito, K. Okano, H-D. Park, T. Itayama, Y. Inamori, B. A. Neilan, B. P. Burns and N. Sugiura, *Microbiology Letters*, 229, 271 (2003).
[47] K. I. Harada, S. Imanishi, H. Kato, M. Mizuno, E. Ito and K. Tsuji, *Toxicon,* 44, 107 (2004).
[48] L. Ho, T. Meyn, A. Keegan, B. Hoefel, C. P. Saint and G. Newcombre, *Water Research*, 40, 768 (2006).
[49] K. Okano, K. Shimizu, Y. Kawauchi, H. Maseda, M. Utsumi, Z. Zhang, B. A. Neilan and N. Sugiura, *Journal of Toxicology,* 1 (2009).
[50] J. Wang, P. Wu, J. Chen and H. Yan, *Biotechnology and Bioengineering Chinese Journal of Chemical Engineering,* 18, 108 (2010).
[51] S. J.Hoeger, B. C. Hitzfeld and D. R. Dietrich, *Toxicology and Applied Pharmacology*, 203: 231 (2005).
[52] H. Yan, J. Wang, J. Chen, W. Wei, H. Wang and H. Wang, *Chemosphere,* 87, 12 (2012).
[53] H. Yan, H. Wang, J. Wang, C. Yin, S. Ma, X. Liu and X. Yin. *Journal of Environmental Sciences,* 24, 1816 (2012).
[54] T. Maruyama, H. D. Park, Z. Ozawa, Y. Tanaka, T. Sumino, K. Hamana, A. Hiraishi and K. Kato, *International Journal of Systematic and Evolutionary Microbiology,* 56, 85 (2006).
[55] G. J. Jones, D. G. Bourne, R. L. Blakeley and H. Doelle, *Natural Toxins*, 2, 228 (1994).
[56] S. Takenaka and M. F. Watanabe, *Chemosphere*, 34, 749 (1997).
[57] Y. H. Kang, C. S. Park and M. S. Han, *Appl. Phycol.,* 24, 1517 (2012).
[58] J. Rapala, K. A. Berg, C. Lyra, R. M. Niemi, W. Manz, S. Suomalainen, L. Paulin and K. Lahti, *Int. J. of Systematic and Evolutionary Microbiology*, 55, 1563 (2005).
[59] L. B. Hu, J. D. Yang, W. Zhou, Y. F. Yin, J. Chen and Z. Q. Chi, *New biotechnology,* 26, 205 (2009).
[60] L. Eleuterio and J. R. Batista, *Toxicon,* 55, 1434 (2010).
[61] J. Chen, J.B. Hu, W. Zhou, S. H. Yan, J. D. Yang, Y. F. Xeu and Z. Q. Shi, *Int. J. Mol. Sci.,* 11, 896 (2010).
[62] A. Ramani, K. Rein, K. G. Shetty and K. Jayachandran, *Biodegradation,* (2012).
[63] P. M. Manage, C. Edwards, B. K. Singh and L. A. Lawton, *Applied and Environmental Microbiology,* 75, 6924 (2009).
[64] K. Ar. Kormas and D. S. Lymperopoulou, *BioMed Research International,* 2013, 1 (2013).
[65] J. Meriluoto, M. Gueimonde, A. Haskard, L. Spoof, O. Sjövall and S. Salminen, *Toxicon,* 46, 111 (2005).
[66] S. M. K. Nybom, S. J. Salminen and J. A. O. Meriluoto, *Toxicon,* 52, 214 (2008).
[67] S. M. K. Nybom, D. Dziga, J. E. Heikkila, T. P. J. Kull, S. J. Salminen and J. A. O. Meriluoto, *Toxicon.,* 59-171 (2012).
[68] L. Hu, F. Zhang, C. Liu and M. Wang, *Energy Procedia*, 16, 2054 (2012).
[69] S. A. Alamni, *Saudi Journal of Biological Sciences,* 19, 435 (2012).
[70] K. Christoffersen, S. Lyck, and A. Winding, *Aquat. Microb. Ecol.,* 27, 125 (2002).

[71] C. Edwards, D. Graham, N. Fowler and L. A. Lawton, *Chemosphere,* 73, 1315 (2008).

[72] L. Giaramida, P. M. Manage, C. Edwards, B. Singh and L. A. Lawton, *International Biodeterioration and Biodegradation,* In press (2012).

[73] R. R. Colwell and D. J Grims DJ. In: R. R. Colwell and D. J. Grimes (Eds.) *ASM Press*, Washington, DC (2000).

[74] T. Maruyama, K. Kato, A. Yokoyama, T. Tanaka, A. Hiraishi and H. D. Park, *Microb. Ecol.*, 46, 279 (2003).

[75] J. Li, K. Shimizu, H. Maseda, Z. Lu, M. Utsumi and Z. Zhang, *Bioresource Technology*, 106, 27 (2012).

[76] A. Kato, S. Y. Imanishi, K. Tsuji and K. Harada, *Water Research,* 41, 1754 (2007).

[77] L. Ho, T. Tang, P. T. Monis and D. Hoefel, *Chemosphere,* 87, 1149 (2012[a]).

[78] S. Merel, M. Clément and O. Thomas, *Toxicon*, 55, 677 (2010).

[79] L. Ho, D. Hoefel, C. P. Saint and G. Newcombe, *Water Research,* 41, 4685(2007).

[80] P. M. Manage, C. Edwards and L. A. Lawton. In: N. Hamamura, S. Suzuki, S. Mendo, C. M. Barroso, H. Iwata and S. Tanabe (Eds), *Terrapub*, 97 (2010).

[81] I. Chorus and L. Mur. In: I. Chorus and J. Bartram (Eds.), *E&FN Spon*, London, 235 (1999).

[82] M. Drikas, G. Newcombe and B. Nicholson, B. *AWWA-WQTC* (2001).

[83] L. Ho, A. L. Gaudieux, S. Fanox, G. Newcombe and A. R. Humpage, *Toxicon,* 50, 438 (2007).

[84] J. Li, K.Shimizu, M.Utsumi, T. Nakamoto, M. K. Sakharkar, Z. Zhang and N. Sugiura, *Journal of Bioscience and Bioengineering*, 111 (6), 695 (2011)

In: Cyanobacteria
Editor: Douglas Davison

ISBN: 978-1-61470-925-1

Chapter 6

MULTIPLE DEFENSE SYSTEMS IN CYANOBACTERIA IN RESPONSE TO SOLAR UV RADIATION

Rajesh P. Rastogi[1,*,#]***, Datta Madamwar***[1,†]
and Aran Incharoensakdi[2,*,‡]
[1]BRD School of Biosciences, Vadtal Road, Satellite Campus,
Sardar Patel University, Anand, Gujarat, India
[2]Laboratory of Cyanobacterial Biotechnology,
Department of Biochemistry, Faculty of Science,
Chulalongkorn University, Bangkok, Thailand

ABSTRACT

The global climate change with an increase in solar ultraviolet (UV) radiations on Earth's surface has generated tremendous concern about its negative impact on all sun-exposed photosynthetic life-forms including cyanobacteria. UV (280 - 400 nm) can affect a number of physiological and biochemical processes in cyanobacteria either directly or indirectly by the generation of reactive oxygen species. Drastic effects caused by UV radiation may imbalance the entire ecosystems. However, many cyanobacteria are able to develop several defense mechanisms against the damaging effects imposed by increased UV radiation. These strategies include the restoration of genetic material by means of excision repair and photoreactivation, repair and re-synthesis of D1 and D2 proteins of PSII reaction center, activation of antioxidant systems, and biosynthesis of several photoprotectants such as mycosporine-like amino acids, scytonemin, carotenoids and polyamines. This chapter presents an overview on the current knowledge of the physiological as well as biochemical adaptation of cyanobacteria with regard to the function of potent defense mechanisms in response to intense solar UV radiation.

[*] Authors for Correspondence: A. Incharoensakdi, Rajesh P. Rastogi.
[#] E-mail: raj_rastogi@rediffmail.com.
[†] E-mail: datta_madamwar@yahoo.com.
[‡] E-mail: aran.i@chula.ac.th.

Keywords: Cyanobacteria, UV radiation, oxidative stress, defense mechanisms, photoprotectants, DNA repair, photoreactivation, excision repair

1. INTRODUCTION

Cyanobacteria or blue green algae are ecologically the most important and dominant component of photoautotrophic microflora in terms of total biomass and productivity in marine, freshwater and terrestrial ecosystems. They are ubiquitous in nature ranging from hot springs to the Arctic and Antarctic regions as well as in other extreme environments of high salinities, pH and irradiances, and may also exist in the form of endosymbionts in plants, lichens and several protists (Whitton 2012). Cyanobacteria are the most primitive forms of Gram-negative prokaryotes and appeared on the Earth between 2.8 and 3.5 x 10^9 years ago during the Precambrian era (Fischer 2008). Due to inherent capacity of photosynthesis mediated oxygen evolution, cyanobacteria are credited for creation of the oxygenic environment which provides a favorable condition for the evolution of existing aerobic life on the Earth's surface (Brocks et al. 1999, Olson 2006).

Cyanobacteria are imperative sources of several industrially important bioactive compounds (Rastogi and Sinha 2009), and also considered as a nonconventional source of several food supplements (Gantar and Svircev 2008). Besides the role of cyanobacteria in agriculture as the potent source of natural biofertilizers, they are also considered as model organisms for space research due to their physiological adaptation and unusual tolerance towards various abiotic factors (Cockell et al. 2011, Olsson-Francis et al. 2013). Cyanobacteria play an important role in biogeochemical cycle of nutrients (Carpenter and Romans 1991), and can maintain the trophic energy dynamics of an ecosystem (Iturriaga and Mitchell 1986, Stock et al. 2014). More recently, cyanobacteria have been considered as an efficient, ecofriendly and alternative source of renewable energy by means of hydrogen (Dutta et al. 2005, Parmar et al. 2011, Khetkorn et al. 2013) and ethanol (Dexter and Fu, 2009; Dienst et al. 2014) production.

In the past few decades, the increase in solar UV radiation on Earth's surface due to depletion of the ozone layer (McKenzie et al. 2007, Manney et al. 2011) has fueled tremendous concern about its negative impact on both aquatic and terrestrial ecosystems (Häder et al. 2011). The solar UV (280-400 nm) radiation can be divided into three groups i.e., ultraviolet A (UV-A, 315-400 nm), ultraviolet B (UV-B, 280-315 nm) and ultraviolet C (UV-C, 100-280 nm). In all the groups of UV radiation, highly energetic UV-C radiation is quantitatively absorbed by oxygen and ozone and lost through atmospheric scattering. Consequently, it does not reach the Earth's atmosphere and hence, shows no effect on the biota of current ecosystems. The wavelength of UV-A radiation can affect the physiology and biochemistry of a cell by indirect photosensitization reaction in the presence of sensitizing molecules such as chlorophylls and phycobilins in algae and cyanobacteria. In contrast, UV-B radiation has less than 1% of total solar radiation reaching the Earth's surface, but is highly active component that brings about chemical modifications in DNA, and changes its molecular structure by the formation of dimers and other DNA lesions due to direct absorption of UV-B radiation by the target molecules (Vincent and Neale 2000, Halliwell and Gutteridge 2007, Rastogi et al. 2010a). Comparison to UV-B radiation, UV-A radiation has poor efficiency in inducing the DNA damage because it is not absorbed by native DNA; however, UV-A radiation can induce DNA damage either by producing a secondary photoreaction of existing DNA photoproducts or by the generation of reactive oxygen species (ROS) via indirect photosensitization reactions as discussed above.

Moreover, prime requirement of solar energy for the key metabolic functions such as photosynthesis and N_2-fixation exposes cyanobacteria simultaneously to harmful doses of UV-A and UV-B radiations in their natural brightly lit habitats. A number of physiological and biochemical processes in cyanobacteria such as growth and survival, motility and orientation, cell differentiation, pigmentation, photosynthetic performance, N_2 metabolism, CO_2 uptake, enzyme activity, genome integrity and total protein profiles have been reported to be affected by UV radiation (Karsten et al. 2007, Sinha et al. 2008, Ma and Gao 2010, Rastogi and Sinha 2011, Rastogi et al. 2011, 2014) (Figure 1). Furthermore, many cyanobacteria have developed multiple defense mechanisms to counteract the damaging effects of UV radiation. These mechanisms include migration, mat formation, synthesis of extracellular polysaccharides (EPS), increased activity of antioxidant systems, biosynthesis of UV-absorbing/screening compounds, repair and resynthesis of D1 and D2 proteins of PSII reaction center, and several DNA repair mechanisms (Figure 1). In this chapter, we have compiled the recent advances and development in the field of UV radiation effects and possible defense mechanisms adopted by cyanobacteria.

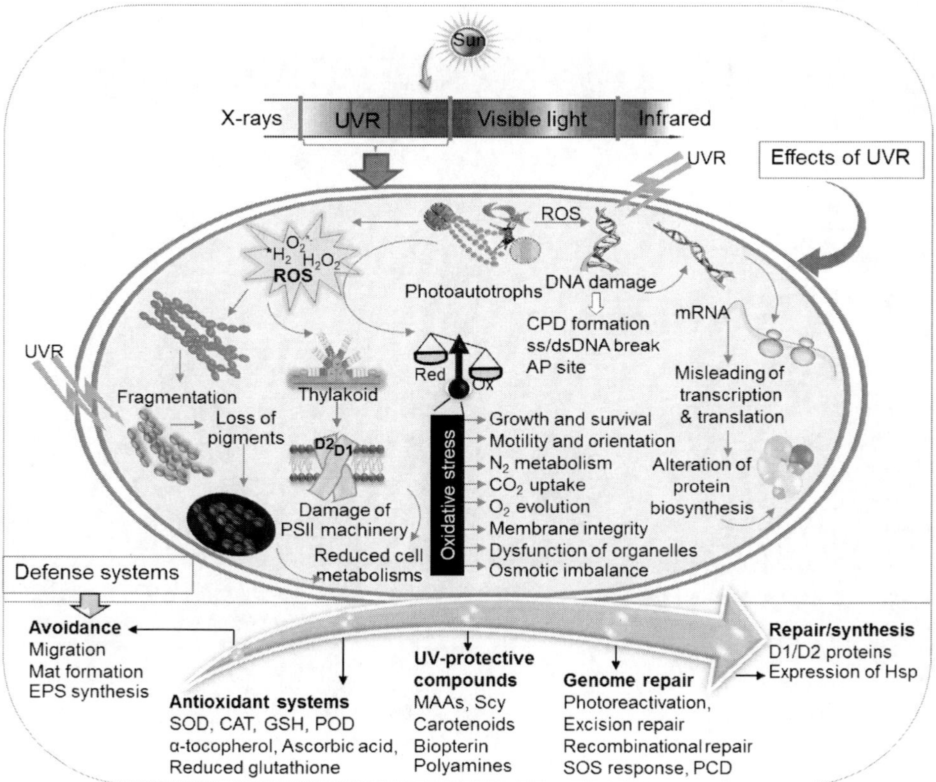

Photo-by Rajesh P. Rastogi.

Figure 1. A diagrammatic view of solar UV radiation-induced effects and adoption of multiple defense systems in cyanobacteria (details in text).

2. Consequences of UV Radiation on General Physiology of Cyanobacteria

UV radiation can affect a number of physiological and biochemical activities of different photosynthetic organisms including cyanobacteria (Sinha et al. 2008). Cyanobacteria exist in both unicellular and filamentous forms with morphologically distinct structures such as vegetative cells, intercalary or terminal heterocysts, akinetes and some other specialized cells (Figure 2). It has been shown that UV-induced oxidative stress affects the cellular organization of cyanobacteria. Morphological changes such as fragmentation of filament or reduced filament length (Wu et al. 2005, Gao et al. 2007, 2008, Ma and Gao 2010, Rastogi et al. 2010b, Singh et al. 2014) and differentiation of vegetative cells into heterocysts and akinetes (Blakefield and Harris 1994) have been reported in different cyanobacteria. It was found that the altered carbon and nitrogen ratio following UV-B exposure could also affect the spacing pattern of heterocysts in the cyanobacterial filament (Sinha et al. 1996). The exact mechanisms of trichome/filament fragmentation or breakage is not known; however, it is assumed that UV-induced higher accumulation of ROS, resulted in the filament breakage by oxidizing the lipids of sheath or cell membrane as well as osmotic imbalance and selective lysis or burst of damaged cells (Ma and Gao, 2010, Rastogi et al. 2010b, Singh et al. 2014) (Figure 3). The filament of *Lyngbya aestuarii* was coiled and differentiated into small bundles in response to UV-B radiation (Rath and Adhikary 2007).

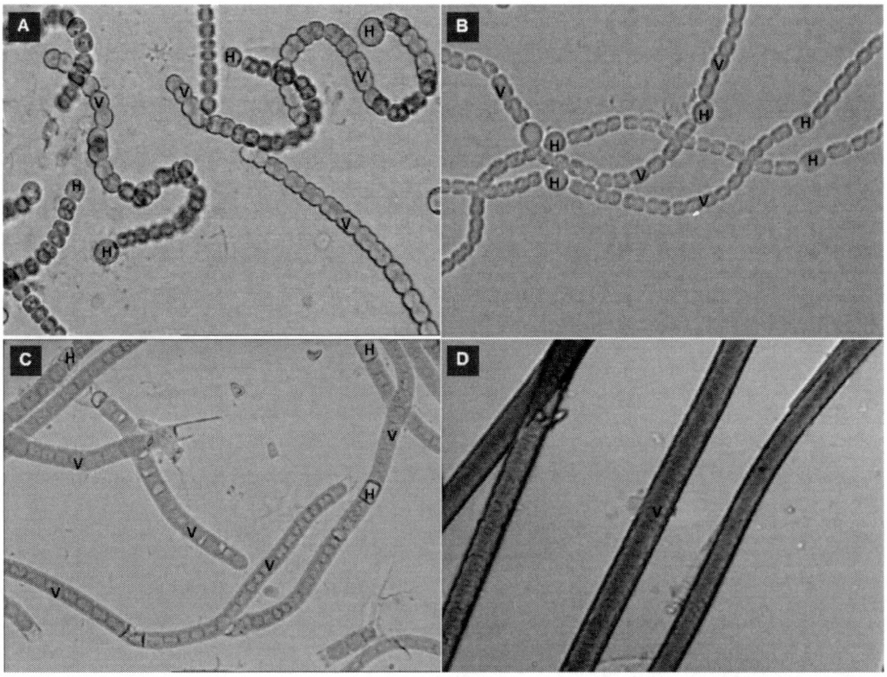

Figure 2. Morphological organization of some heterocystous cyanobacteria such as *Nostoc* sp. (A), *Anabaena* sp. (B) and *Scytonema* sp. (C), and non-heterocystous cyanobacterial species such as *Lyngbya* sp. (D). H- heterocyst, V-vegetative cells.

UV radiation severely affects the motility and orientation mechanisms (Castenholz 1982, Häder 1984, Donkor et al. 1993, Kruschel and Castenholz 1998) and impedes the required adaptation of cyanobacteria against day-to-day fluctuating environmental conditions with high light intensity. UV radiation impairs the linear velocity of the cyanobacterial cells with subsequent loss of their ability to escape from intense solar radiation (Donkor and Häder 1991, Donkor et al. 1993). Kruschel and Castenholz (1998) have shown the effects of visible and solar UV radiation on the vertical movements of *Oscillatoria laetevirens* and *Spirulina subsalsa*. A significant decrease in percentage of motility was observed in several cyanobacterial filaments under intense solar light with UV radiation. Interestingly, removal of UV radiation significantly increased their motility, indicating the potential negative impact of UV radiation on the gliding movement of cyanobacteria (Donkor et al. 1993). It was observed that inhibition of phototactic orientation and photophobic responses reduced the ability of organisms to orient in their intense photo environment (Häder and Worrest 1991). The rate of gliding movement in response to UV radiation may vary among different species/strains of cyanobacteria (Quesada and Vincent 1997). Moreover, different gene products have been reported to be involved in phototactic movement in the unicellular cyanobacterium *Synechocystis* PCC 6803 (Bhaya et al. 2000, Bhaya 2004, Yoshihara and Ikeuchi 2004).

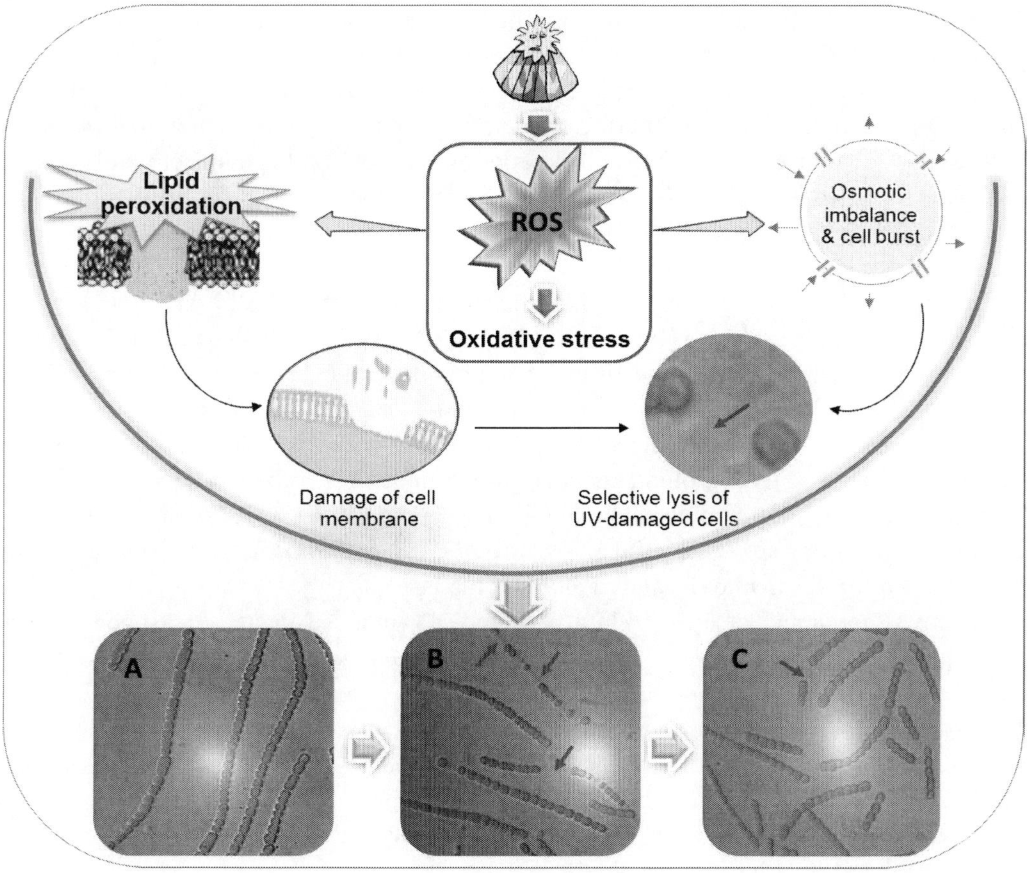

Figure 3. Schematic presentation of possible mechanisms of UV radiation-induced filament fragmentation in cyanobacteria (details in text).

Increased UV radiation exerts deleterious effects on growth and survival of cyanobacteria (Sinha et al. 1995a). In one study, complete killing and loss of survival was observed in cyanobacteria within 2-3 h of UV-B exposure (Sinha et al. 1995a). Moreover, exposure of *Oscillatoria priestleyi* and *Phormidium murrayi* to the same dose of UV-B radiation resulted in a decrease in growth by 100% and 62 %, respectively, indicating that several species of cyanobacteria show differential sensitivity to UV radiation in terms of their certain metabolic processes including growth and survival (Tyagi et al. 1992, Quesada and Vincent 1997). Similarly, a differential reduction in growth by 32, 88 and 95 % was observed in *Nostoc muscorum*, *Plectonema boryanum* and *Aphanothece* sp., respectively after 30 min of UV-B exposure (0.4 Wm^{-2}) for 10 days (Zeeshan and Prasad 2009). Recently, Rastogi et al. (2014) also observed a marked reduction in growth of *Rivularia* sp. HKAR-4 under UV radiation.

The growth of cyanobacteria is associated with photosynthetic pigment and rate of photosynthesis. It has been shown that UV-B radiation causes photobleaching of photosynthetic pigments such as chlorophyll *a* (chl *a*) and phycobiliproteins in many cyanobacteria (Sinha et al. 1995b, Sinha and Häder 1998, Prasad and Zeeshan 2004). The levels of photosynthetic pigments such as chl *a*, myxoxanthophyll, and β-carotene was decreased by 74, 81 and 86 % , respectively after 6 h UV-B irradiation (0.8 ± 0.1 mW cm^{-2}) in the marine cyanobacterium *Phormidium tenue* (Bhandari and Sharma 2011). Moreover, high intensity of PAR and UV radiation delays the transfer of excitation energy to photosystems, and also damages the photosynthetic machinery (Campbell et al. 1998, Sinha and Häder 2003) leading to a decrease in growth and photosynthetic performance of the organism (Bhandari and Sharma 2006). Complete bleaching of photosynthetic pigments was observed after 2-3 h of UV-B (5 Wm^{-2}) irradiation in the rice field cyanobacterium, *Aulosira fertilissima* (Banerjee and Häder 1996). A rapid destruction of α-PC, β- PC and both α and β –APC was observed under UV-B (1.3 Wm^{-2}) radiation in *Synechocystis* PCC 6803 (Rinalducci et al. 2006).

Distortion of thylakoid membrane with reduced chl *a* content was observed under UV-B radiation in *Spirulina platensis* (Gupta et al. 2008). UV-induced degradation of D1 and D2 proteins of PS-II reaction centers has been reported in several cyanobacteria (Campbell et al. 1998, Six et al. 2007, Vass 2012). Moreover, any drastic changes in photosynthetic machinery as discussed above can influence the rate of photosynthesis. Decreases in optimum photosynthetic yields (*Fv*/*Fm*), pigments, and growth under UV radiation have been reported in several cyanobacterial strains such as *Anabaena* sp. PCC 7120 (Gao et al. 2007), *Arthrospira* (*Spirulina*) sp. (Gao et al. 2008), *Anabaena variabilis* PCC 7937 (Singh et al. 2013a) and *Phormidium tenue* (Bhandari and Sharma 2011).

Recently, Giordanino et al. (2011) also found UV-induced decrease in photosynthetic yield in the four strains of cyanobacteria such as *Anabaena* sp., *Nostoc* sp., *Arthrospira platensis* and *Microcystis* sp. Furthermore, several studies have revealed the harmful effects of UV radiation on photosynthetic performance and its recovery after long term exposure to mild PAR or UV radiation (Karsten et al. 2007). The photosynthetic yield of *A. variabilis* PCC 7937 was severely inhibited immediately after 10 min of high PAR and UV radiation; however, the recovery of photosynthetic yield was observed after 2 h of the exposure, which continued for the next 24 h (Singh et al. 2013a).

The UV radiation (mainly UV-B) also affects the process of N_2-fixation by affecting the primary N_2-fixing enzyme, nitrogenase (Kumar et al. 1996, Lesser 2008). A complete loss of nitrogenase activity was observed in several rice-field cyanobacteria such as *Anabaena* sp.,

Nostoc sp., *Calothrix* sp. and *Scytonema* sp. after 25-40 min of UV-B irradiation (Kumar et al. 2003).

Besides nitrogenase, the nitrate reductase activity has also been observed to be affected by UV radiation (Rai et al. 1998). Moreover, about 80 % annual input of nitrogen takes place via biological nitrogen fixation by cyanobacteria (Solheim et al. 2006), and any damaging effects of UV radiation on a nitrogenase may affect the total annual input of nitrogen in an ecosystem (Solheim et al. 2006). Moreover, besides the above mentioned effects, UV radiation also affects several other parameters such as CO_2 uptake, O_2 evolution, ribulose-1,5 bisphosphate carboxylase (RuBISCO) activity, ATP and NADPH supply in cyanobacteria (Sinha et al. 2008).

3. UV-Induced Generation of ROS

The high-energetic UV-B radiation may affect normal state of life either through direct effects on cellular DNA and proteins or indirectly by the production of ROS. The ROS such as singlet oxygen ($^1O_2$), superoxide anion (O_2^-), hydrogen peroxide (H_2O_2) and the hydroxyl radical ($OH^·$) are potent and highly active oxidizing agents. In photosynthetic organisms such as cyanobacteria, ROS are produced by means of photosynthetic electron transport chain (Figure 4) (Latifi et al. 2009). Production of ROS can also be seen as a result of normal metabolic activity.

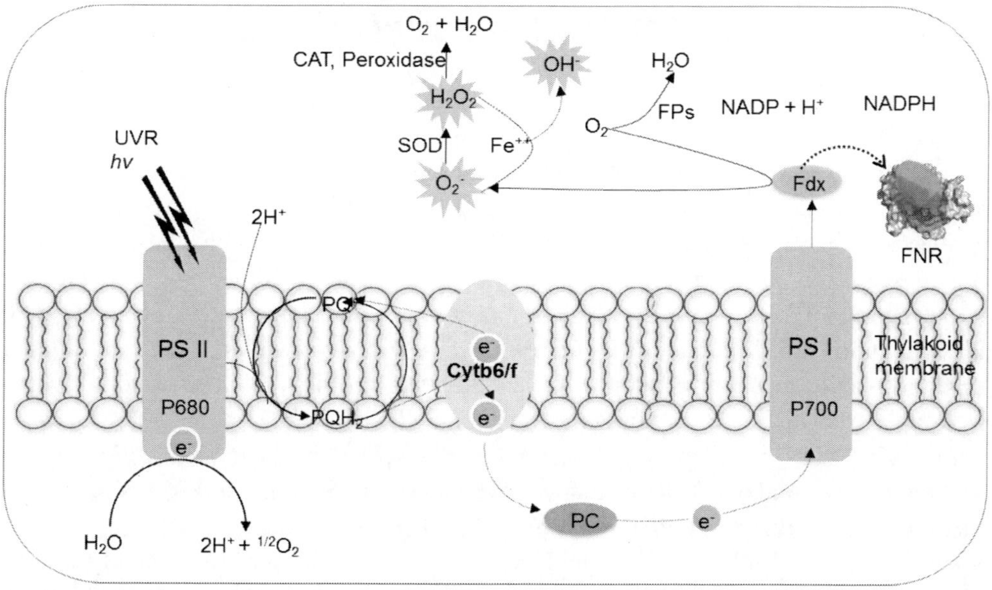

For details see Latifi et al., 2009.

Figure 4. A simplified diagrammatic view of ROS generation *via* photosynthetic electron transport chain in PSI and PSII. Reduction of O_2 by means of electrons relocating from PSII results in production of O_2^- radicals in PSI. In the presence of SOD, O_2^- radical forms H_2O_2 and subsequently, the H_2O_2 is reduced to form H_2O in the presence of CAT and peroxidases. The reduction of H_2O_2 by metal catalyzed Fenton reaction produces OH^- radicals. FNR (Fdx + NADP + reductase) catalyzes one terminal step of the conversion of light energy into chemical energy during photosynthesis.

It has been observed that intense light beyond the normal capacity of the photosynthetic electron flow may cause production of additional ROS along with the $^1O_2$ leading to inactivation of photosystems (Mehler 1951). ROS are unavoidably generated as intermediates of O_2 reduction, or by its energization. Contrary to higher plants and algae, cyanobacteria undergo a high degree of O_2 reduction by consuming 50 % of the photosynthetic electrons compared to only 15 % in plants (Badger et al. 2000). However, production of H_2O_2 in the cyanobacterium *Synechocystis* PCC 6803 stands for 1 % of the maximum rate of photosynthetic electron transport *in vivo* (Tichy and Vermaas 1999). Moreover, photosynthetic production of H_2O_2 has also been reported in *Anacystis nidulans* (Patterson and Myers 1973).

UV-induced production of ROS has been reported in some cyanobacteria (He and Häder 2002a, Rastogi et al. 2010b). Recently, Rastogi et al. (2010b) for the first time detected UV-induced generation of ROS by confocal image analysis in *Anabaena variabilis* PCC7937 using a ROS sensing probe 2', 7'-dichlorodihydrofluorescein diacetate (DCFH-DA). Very little is known about the production of ROS and subsequent effects of oxidative stress in cyanobacteria (Nishiyama et al. 2006, Latifi et al. 2009). More extensive study is needed to elucidate the molecular mechanisms of ROS production and its biological effects on cyanobacteria. In the subsequent section, we are considering the UV-induced ROS mediated effects as well as a general and broad effect of UV radiation on cyanobacteria.

4. UV-Induced Proteins and Genomic Instability in Cyanobacteria

UV radiation can damage several important biomolecules such as lipids, proteins and DNA of the cell either directly or by means of ROS (Rastogi et al. 2010a). Proteins and DNA are primary targets of UV radiation due to their prominent absorbance in the UV-B (280-315 nm) range. UV-induced degradation of proteins has been reported in several cyanobacteria. A complete loss of total cell protein bands was observed after 150 min UV-B exposure in *Nostoc commune* and *Scytonema* sp. (Sinha et al. 1995a, 1995b). Similarly, complete loss of all protein bands after 120 min of UV-B exposure was also apparent in *Nostoc calcicola* (Kumar et al.1996). UV irradiated cells of *Lyngbya aestuarii* showed repression of 20 and 22 kDa proteins; however, overproduction of 84, 73, 60, 46, 40, 37 kDa proteins was also observed after 6-24 h UV-B irradiation (Rath and Adhikari 2007). Qualitative as well as quantitative changes in total protein profile were observed in *Nostoc spongiaeforme* and *Phormidium corium* under UV-B irradiation (Bhandari and Sharma 2006). UV exposure of *Synechococcus* sp. WH8102 resulted in a loss of D1 (PsbA) proteins (Six et al. 2007).

Besides proteins, lipids and fatty acids are also structurally and functionally important biomolecules naturally occurring in several organisms including cyanobacteria. They play a significant role in the tolerance to several physiological stressors such as heat and UV radiation (Singh et al. 2002). UV radiation can affect the lipid moiety and integrity of the cell. UV-induced lipid peroxidation in terms of increased malondialdehyde (MDA) content has been reported in several cyanobacteria such as *Anabaena* sp. (He and Häder, 2002b), *Cylindrospermum* sp. (Chris et al. 2006), *P. corium* (Bhandari and Sharma, 2006), *Nostoc muscorum*, *Plectonema boryanum*, *Aphanothece* sp. (Zeeshan and Prasad 2009), and

Synechocystis PCC 6803 (Jantaro et al. 2014). UV-induced loss of saturated fatty acid (SFA), monounsaturated fatty acid (MUFA) and polyunsaturated fatty acid (PUFA) was observed in *S. platensis* (Gupta et al. 2008). A decrease in SFA was also observed in *N. spongiaeforme* in response to UV-B radiation (Bhandari and Sharma 2006).

UV-induced ROS can damage the components of DNA such as purine/pyrimidine bases and deoxyribose sugar backbone leading to single/double DNA strand breaks. It has been established that exposure of DNA to UV radiation led to the formation of a number of oxidation products of purine such as 8-oxo-7,8-dihydroguanyl (8-oxoGua), 8-oxo-Ade, 2,6-diamino-4-hydroxy-5-formamidoguanine (FapyGua), FapyAde, and oxazolone (Doetsch et al. 1995, Hall et al. 1996). Moreover, cyclobutane pyrimidine dimers (CPDs) and pyrimidine 6-4 pyrimidone photoproducts (6-4PPs) are the major DNA lesions formed by UV-B irradiation (Figure 5).

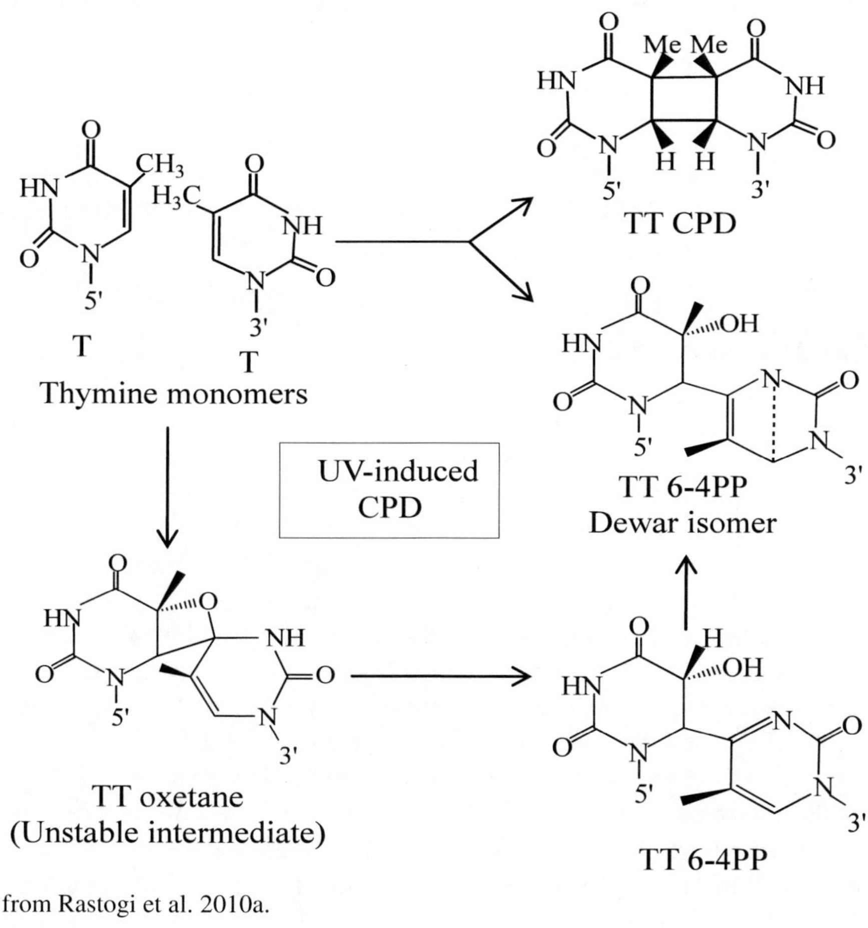

Modified from Rastogi et al. 2010a.

Figure 5. UV-induced formation of thymine-thymine CPD, 6-4PP and their Dewar valence isomer.

It has been found that the DNA lesion 6-4PPs are readily converted into their Dewar valence isomers upon exposure to UV-A/B radiation (Taylor et al. 1990, Rastogi et al. 2010a). UV-induced DNA damage has been reported in some cyanobacteria (O'Brien and Houghton 1982a, Blakefield and Harris 1994, He and Häder 2002b, Kumar et al. 2004,

Rastogi et al. 2011, Rastogi and Sinha 2011, Wang et al. 2012, Rastogi et al. 2014). UV-induced formation of thymine dimers (T<>T CPD) has been observed in several cyanobacteria such as *Anabaena* sp., *Nostoc* sp., *Scytonema* sp. and *Rivularia* sp. (Sinha et al. 2001, Rastogi et al. 2011). UV-induced formation of CPDs has also been reported in the economically important cyanobacterium *A. platensis* (Gao et al. 2008). UV radiation-induced delay in chromosome replication was detected in a marine cyanobacterium *Prochlorococcus marinus* PCC9511 (Kolowrat et al. 2010). Recently, Rastogi et al. (2014) have reported UV-induced T<>T CPD in *Rivularia* sp. HKAR-4. To get more information regarding the molecular mechanisms of UV-induced DNA damage and their analysis, readers are suggested to see the paper by Rastogi et al. (2010a).

5. UV-Induced Defense Mechanisms in Cyanobacteria

Cyanobacteria are one of the most dominant photoautotrophs on the planet, and any adverse effects on their existence may lead to an imbalance of entire ecosystem. However, several cyanobacteria have developed multiple defense systems in response to adverse effects of UV radiation (Figure 1) (Castenholz, 1997, Sinha et al. 2008). In the subsequent sections, we discuss some important defense mechanisms reported in some cyanobacterial strains against high energetic solar UV radiations reaching their natural habitats.

5.1. Primary Preventive Mechanisms

Cyanobacteria have a very long life history connected with earlier and present UV radiation stress. During the course of evolution they have evolved various defense mechanisms including some photokinetic or photophobic reactions such as avoidance/escape by means of migration or mat formation, and morphological transformation as a primary preventive mechanism against the damaging effects of enhanced UV radiation. In general, cyanobacteria can avoid or escape from high solar radiation by downward migration and sinking deeper into the water column (Reynolds et al. 1987) (Figure 6). The incidence of daily fluctuating light intensity affects the upward and/or downward migration patterns of cyanobacteria in their natural habitats (Richardson and Castenholz 1987, Kruschel and Castenholz 1998). Kruschel and Castenholz (1998) found that *Oscillatoria laetevirens* and *Spirulina subsalsa* protect their photosynthetic machinery by downward migration by means of vertical gliding movement. Similarly, UV-induced upward and downward migration was observed in *Oscillatoria* sp. inhabiting a microbial mat (Nadeau et al. 1999). The migratory behavior of motile Oscillatoriales into the mat matrix was also observed during intense light condition (Ramsing and Prufert-Bebout 1994). *Microcoleus chthonoplastes* in the microbial mats of lake Sinai, Egypt showed highest incidence of migration in response to UV-B radiation in comparision to UV-A and PAR (Bebout and Garcia-Pichel 1995). Overall, the upward or downward and lateral migration of cyanobacterial cells/trichomes probably depends on the intensity of solar radiation reaching their natural habitats (Ramsing and Prufert-Bebout 1994). Besides migration, some species/strains of cyanobacteria form a microbial mat to avoid or minimize the harmful effects of intense solar light and UV radiation

(Quesada and Vincent 1997, Büdel 1999). Both aquatic and terrestrial cyanobacteria often form dark green or black coloured crusts at solid-air interface under a wide range of environmental conditions. Moreover, terrestrial habitats are more favourable for mat formation than aquatic systems due to periodic long-term desiccation and intense solar radiation (Potts 1999).

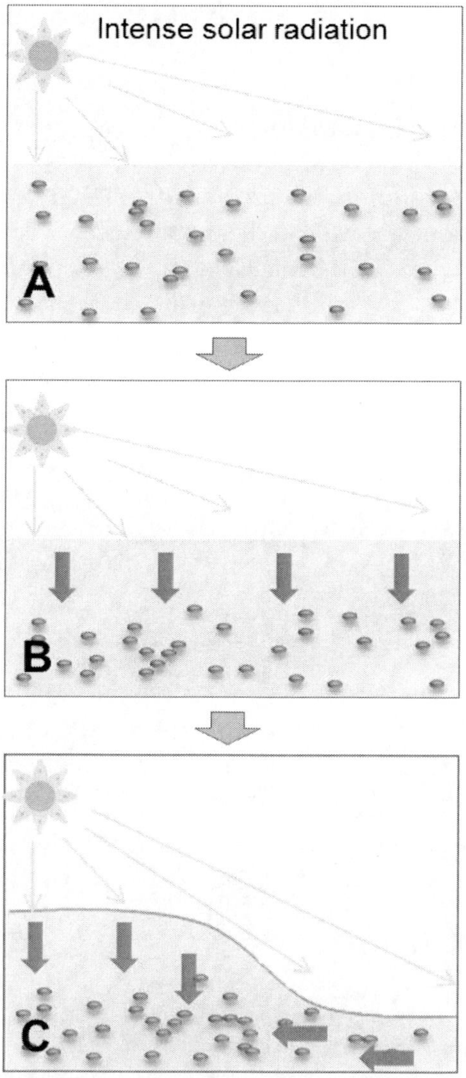

Figure 6. Diagrammatic presentation of incidence of intense solar radiation (A) and its avoidance by means of downward (B) and/or lateral (C) migration or sinking deep into water column of cyanobacteria in an aquatic ecosystem.

The formation of microbial mats and their composition as well as consistency depends on the microbial communities and environment of a particular habitat. Contrary to unicellular cyanobacteria, filamentous cyanobacteria often form mat-like structure due to high deposition of extracellular sheath compounds (Tomaselli and Giovannetti 1993, Sutherland 2009).

It has been shown that some cyanobacterial species protect themselves from UV radiation by adopting the self-shading effects. Self-shading as an effective protective mechanism against UV radiation was observed in *Arthrospira platensis*, where helix pitch of spiral structure of this cyanobacterium was decreased and found to be more compact (Wu et al. 2005). Overall, the primary preventive mechanisms as discussed above can be considered as the first-line of defense mechanisms adopted by cyanobacteria against damaging effects of UV radiation.

5.2. Biochemical Defense Mechanisms

5.2.1. Biosynthesis of Extracellular Polysaccharides (EPS)

The EPS are high molecular weight heteropolysaccharide compounds produced by different cyanobacteria (Figure 7) (Nicolaus et al. 1999, Shah et al. 2000, Parikh and Madamwar 2006) at different stages of their growth. They serve as a boundary between the cells and the surrounding environment and play an important role in mitigation strategy against desiccation and harmful effects of UV radiation (De Caiola et al. 1996, Ehling-Schulz et al. 1997, Chen et al. 2009).

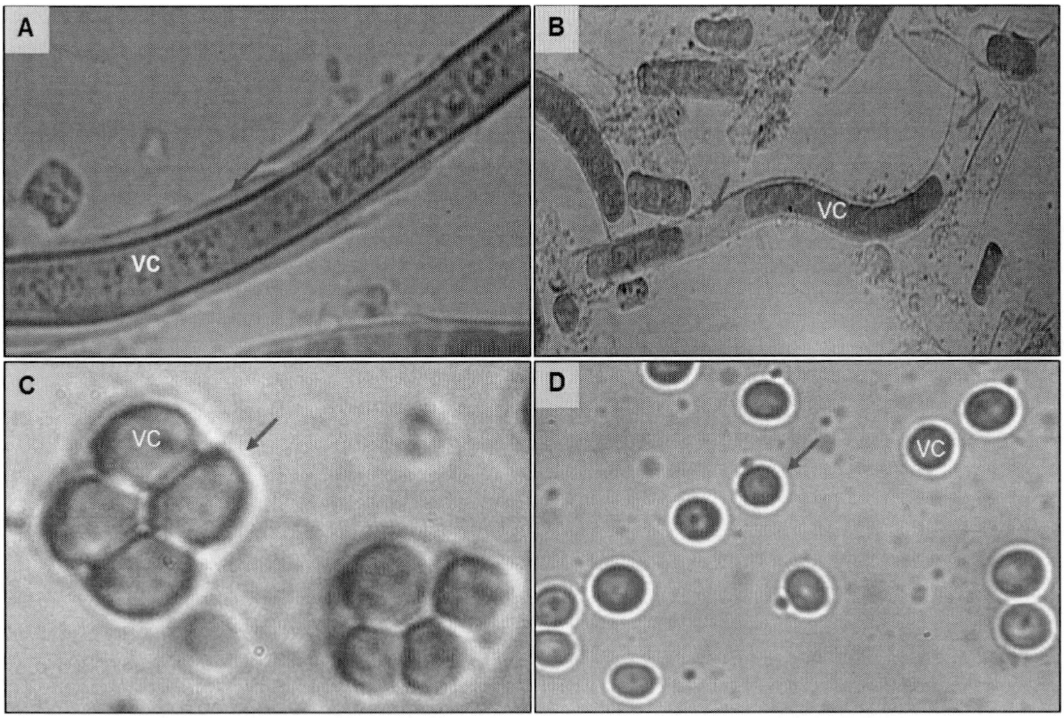

Figure 7. The EPS (shown by red arrow) produced by different cyanobacterial species.

In general, EPS are characterized by an extreme structural diversity and play diverse role in nature including protection from desiccation and UV stress. The photoprotective function of EPS depends on their thickness and consistency. It has been shown that the microorganisms having very dense EPS are relatively highly resistant to desiccation and high solar radiation. Biosynthesis or accumulation of EPS under UV-B radiation strongly supports their UV-protective role in cyanobacteria (Ehling-Schulz et al. 1997, Quesada et al. 1999). EPS exhibited a significant protective effect on DNA strand breaks and lipid peroxidation by reducing the UV-B-induced production of ROS in a desert cyanobacterium *Microcoleus vaginatus* (Chen et al. 2009). Similarly, Li et al. (2011) revealed a free radical scavenging activity of the EPS from *Nostoc commune*. A number of environmental factors affect the biosynthesis of EPS (Pereira et al. 2009); however, light is one of the most important factors that increases the synthesis of EPS (Yu et al. 2010, Mota et al. 2013, Ge et al. 2014, Han et al. 2014).

5.2.2. Antioxidant Defense Systems

The occurrence of antioxidant defense systems may protect the cyanobacterial cells by scavenging the excessive ROS produced under enhanced UV radiation (He and Häder 2002a). Cyanobacteria have devised some complex antioxidant enzymatic or non-enzymatic systems to prevail over the UV-induced oxidative damage (Singh et al. 2013b). A number of enzymatic antioxidants such as ascorbate peroxidase (APX), catalase (CAT), dehydroascorbate reductase (DHAR), glutathione reductase (GR), glutathione peroxidase (GPX), and superoxide dismutase (SOD), as well as non-enzymatic antioxidants such as ascorbic acid (ASA; vitamin C), carotenoids, α-tocopherols (α-TOC; vitamin E) and reduced glutathione have been reported in different photosynthesizing organisms including cyanobacteria (Ishikawa and Shigeoka 2008, Bernroitner et al. 2009, Singh et al. 2010, Singh et al. 2013b). The enzymatic antioxidants APX and CAT play an important role in scavenging of H_2O_2 in cyanobacteria (Miyake et al. 1991, Tichy and Verma 1999); however, in some cyanobacteria the APX is not found and CAT alone acts as H_2O_2 scavenger (Mutsuda et al. 1996). Moreover, CAT-orthologues have been reported to be present in 20 cyanobacterial genomes (Latifi et al. 2009). The SOD is considered a key primary antioxidant molecule which exists in different metalloforms such as Fe-SOD, Mn-SOD, Cu/Zn-SOD and Ni-SOD and protects the cellular system from oxidative stress by detoxifying the toxic superoxide radicals (Lesser and Stochaj 1990, Fridovich 1997, Ismaiel et al. 2014, Ke et al. 2014). SOD removes superoxide radicals (O_2^-) and converts them to H_2O_2 which is further converted into H_2O and O_2 through a combined catalase-peroxidase system (Tel-Or et al. 1986). Moreover, different metalloforms of SOD have been reported in a number of cyanobacterial species (Wolfe-Simon et al. 2005, Priya et al. 2007, 2010). An increased activity of Fe-SOD was observed under desiccation stress and rehydration during UV-A and UV-B irradiation in *N. commune* DRH1 (Shirkly et al. 2000, Ehling-Schulz et al. 2002). Increased synthesis of Mn-SOD was also observed against oxidative stress in *Anabaena* PCC 7120 (Zhao et al. 2007). Induction of SOD and APX as well as some non-enzymatic compounds such as ASA, carotenoids and α-TOC was observed under UV-B radiation in *Anabaena doliolum* (Bhargava et al. 2007). Cameron and Pakrasi (2010) reported that glutathione can play a role in acclimation to environmental and redox perturbations in *Synechocystis* PCC 6803. It has been shown that GPX plays a significant role in removal of lipid hydroperoxides under normal and stress conditions in *Synechocystis* PCC 6803 (Gaber et al. 2004). Glutathione reductase (GR) was also found to act as peroxide scavenger which protected *Nostoc muscorum* against oxidative stress (Karni et al. 1984). Besides the role of antioxidant enzymatic defense systems, several non-enzymatic antioxidants have also been reported in cyanobacteria. Wang et al. (2008) have reported the involvement of the antioxidant systems including exogenous N-acetylcysteine and ascorbic acid in protecting *Nostoc* sp. against UV-B radiation.

5.2.3. Biosyntheis of Photoprotective Compounds

Biosynthesis of some UV-absorbing/screening compounds such as mycosporine-like amino acids (MAAs) and scytonemin plays an important role in photoprotection of several taxonomic groups including cyanobacteria (Rastogi et al. 2010c). MAAs are small, colourless, water-soluble compounds composed of cyclohexenone or cyclohexenimine chromophores conjugated with the nitrogen substituent of an amino acid or its imino alcohol. More than 24 MAAs have been reported in nature; however, mycosporine-glycine, mycosporine-2-glycine, shinorine, porphyra-334, palythinol, palythene, asterina-330 and euhalothece-362 (Figure 8) are the most common MAAs reported in different cyanobacteria. Strong UV-absorption maxima (310 to 362 nm), high molar extinction coefficients (ε = 28,100–50,000 M^{-1} cm^{-1}), photostability in both fresh and saline water in the presence of photosensitizers, ability to prevent UV-induced skin damage, antioxidant properties, resistant to several abiotic stressors such as temperature, UV radiation, various solvents and pH confers strong support in favour of MAAs as photoprotective compounds (Dunlap and Yamamoto 1995, Conde et al. 2000, Gröniger and Häder 2000, Whitehead and Hedges 2005, de la Coba et al. 2009, Rastogi and Incharoensakdi 2014a, 2014b, 2014c). Recently, a number of different MAAs have been reported in different cyanobacterial strains (Rastogi et al. 2012, Rastogi and Incharoensakdi 2014a, 2014b, 2014c). A number of abiotic factors such as PAR and UV radiation, nutrients, osmotic and ionic stress, desiccation and light/dark condition have been reported to affect the biosynthesis of several MAAs in cyanobacteria (Rastogi et al. 2010c). MAAs can be highly induced by PAR and UV-A/B irradiation (Rastogi et al. 2014a, 2014b, 2014c).

In one study, circadian induction of MAA under alternate light (PAR+UVR) and dark condition was observed in *Scytonema* sp. HKAR-3 (Rastogi et al. 2010d), indicating that biosynthesis of MAA is an energy dependent process. It has been found that MAA can prevent 3 out of 10 photons from hitting cytoplasmic targets in cyanobacteria (Garcia-Pichel et al. 1993), and can dissipate absorbed radiation as heat without producing reactive oxygen ROS (Conde et al. 2000, 2004). Several MAAs have been found to act as strong antioxidants (Rastogi and Incharoensakdi 2014a, 2014c). MAA can also block the formation of UV-induced thymine dimers (Misonou et al. 2003). The biosynthesis of MAA involved a shikimate pathway intermediate (Balskus and Walsh 2010). A cluster of four genes was found in *A. variabilis* responsible for the conversion of the common pentose phosphate pathway (PPP) intermediate sedoheptulose-7-phosphate (S7-P) into shinorine via 4-deoxygadusol (Balskus and Walsh 2010) (Figure 8).

It was found that a dehydroquinate synthase (DHQS) homolog Ava_3858 and O-methyltransferase (O-MT) Ava_3857 convert the precursor into 4-deoxygadusol and subsequently adenosine triphosphate (ATP)-grasp homolog Ava_3856 converts the 4-deoxygadusol and glycine into mycosporine-glycine. The nonribosomal peptide synthetases (NRPS)-like enzyme Ava_3855 attaches mycosporine-glycine to serine to form shinorine. Moreover, the exact biosynthetic route of all MAAs is not yet well known.

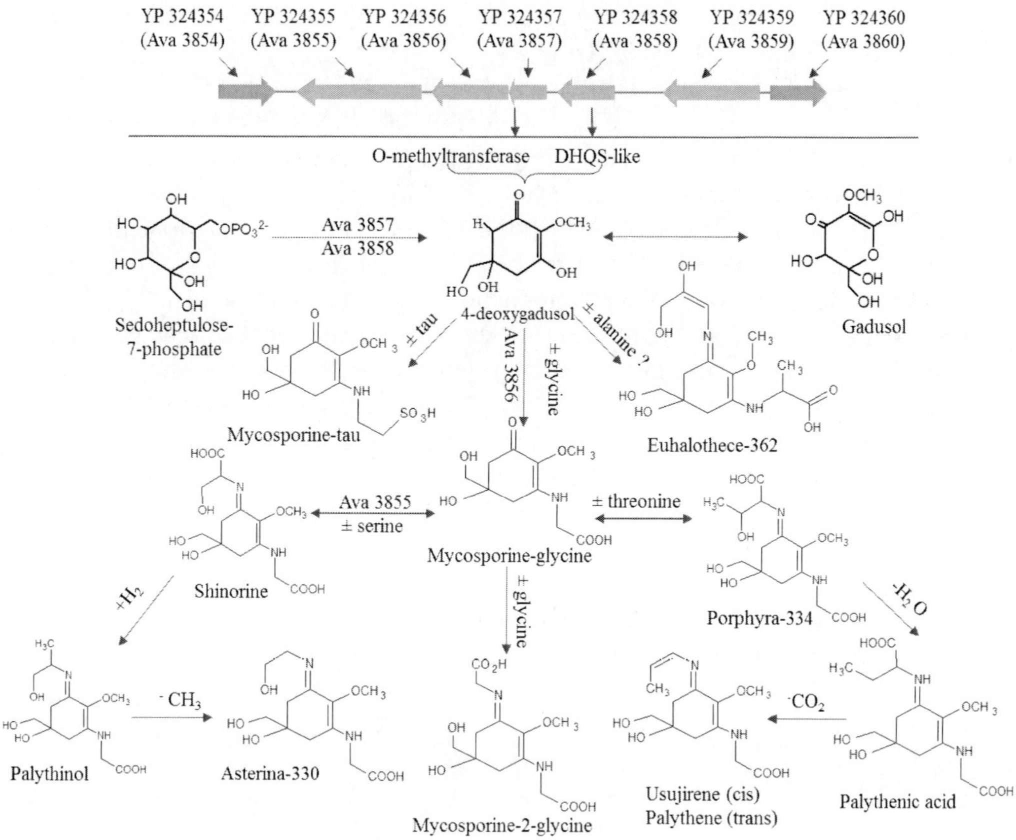

Adapted from Balskus and Walsh 2010, Rastogi et al. 2010c.

Figure 8. Proposed gene cluster for MAAs biosynthesis as reported in the cyanobacterium *Anabaena variabilis* PCC7937 and possible interconversion of some common MAAs found in cyanobacteria.

Besides MAAs, scytonemin is also an important pigment reportedly found in diverse strains of cyanobacteria (Rastogi et al. 2010c, Rastogi et al. 2013, Rastogi and Incharoensakdi 2014b). It is a lipid-soluble dimeric compound composed of indolic and phenolic subunits (Proteau et al. 1993) and exists in both oxidized (MW 544 Da) and reduced (MW 546 Da) forms (Squier et al. 2004, Rastogi and Incharoensakdi 2014b). Scytonemin is located in the extracellular polysaccharide sheath of some cyanobacteria and acts as a passive sunscreen in photoprotection against UV radiation (Garcia-Pichel and Castenholz 1991, Garcia-Pichel et al. 1992). The UV absorption maximum of scytonemin is at 386 nm; however, considerable absorption can also be detected at 251, 278 and 300 nm (Figure 9). Scytonemin can reduce about 90 % solar UV-A radiation that can reach the cell (Garcia-Pichel and Castenholz 1991, Garcia-Pichel et al. 1992). It is a highly stable pigment under different abiotic stresses (Rastogi and Incharoensakdi 2014b) and performs its UV screening function without any further metabolic investment (Fleming and Castenholz 2007). Three new derivatives of scytonemin pigment such as dimethoxyscytonemin, tetramethoxyscytonemin and scytonin (Figure 10) have been derived from the organic extracts of *Scytonema* sp. (Bultel-Poncé et al. 2004). Recently, a new pigment (based on scytoneman skeleton), scytonemin-3a-imine

(Figure 10) with UV/Vis absorption maxima at 237, 366, 437 and 564 nm was isolated from *Scytonema hoffmani* (Grant and Louda 2013).

The biosynthesis of scytonemin is affected by several environmental factors (Dillon et al. 2002, Rastogi et al. 2013, Rastogi and Incharoensakdi 2014b). It has been shown that high photon fluence of PAR and UV-A radiation, as well as temperature and oxidative stress in combination with UV-A strongly induce scytonemin biosynthesis in cyanobacteria (Garcia-Pichel and Castenholz 1991, Dillon et al. 2002). The biosynthesis of scytonemin probably involves tryptophan and tyrosine derivatives that absorb ambient UV radiation (Proteau et al. 1993). A gene cluster responsible for the production of scytonemin was identified in *Nostoc punctiforme* ATCC 29133 (Soule et al. 2007). Recently, several cyanobacterial genomes have been examined to identify the scytonemin biosynthetic gene cluster across a number of cyanobacterial lineages (Balskus and Walsh 2008, 2009, Sorrels et al. 2009, Soule et al. 2009, Balskus et al. 2011). The scytonemin gene cluster consists of unidirectionally transcribed 18 open reading frames (ORFs: NpR1276–NpR1259) including a total of eight genes involved in the biosynthesis of tryptophan and tyrosine while other genes do not show any significant homology with functionally characterized proteins (Rastogi et al. 2010c). Moreover, some genetic variation exists between genome clusters, but the majority of the scytonemin synthesizing genes show high degree of amino acid sequence similarity, indicating the highly conserved process for scytonemin biosynthesis in cyanobacteria (Soule et al. 2009, Balskus et al. 2011).

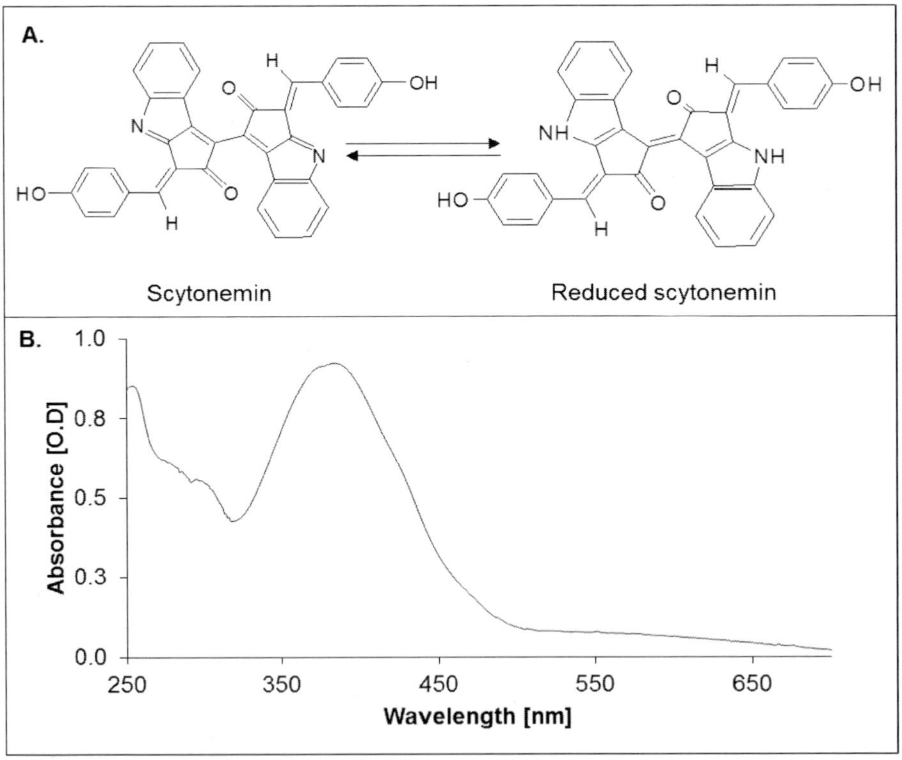

Adapted from Rastogi and Incharoensakdi, 2014b.

Figure 9. Chemical structure of scytonemin found in both oxidized (A) and reduced form (B) with UV-absorption maximum at 386 nm.

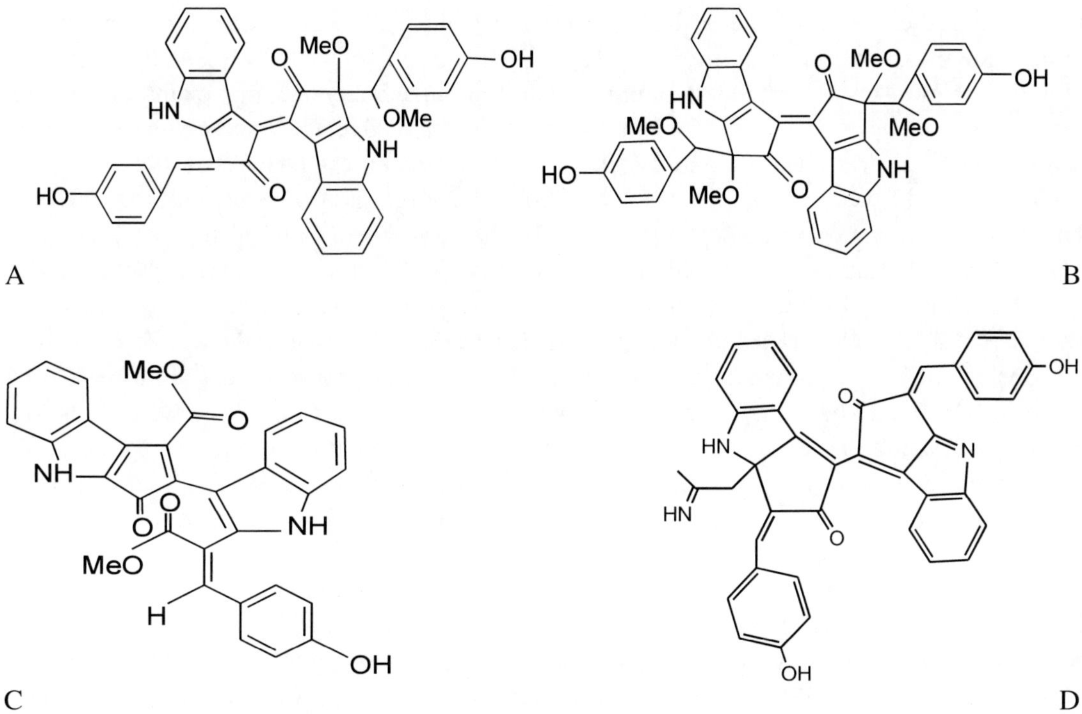

Modified from Garcia-Pichel and Castenholz 1991, Bultel-Poncé et al. 2004, Grant and Louda 2013.

Figure 10. Structure of scytonemin derivatives such as dimethoxyscytonemin (A), tetramethoxyscytonemin (B), scytonin (C) and scytonemin-3a-imine (D).

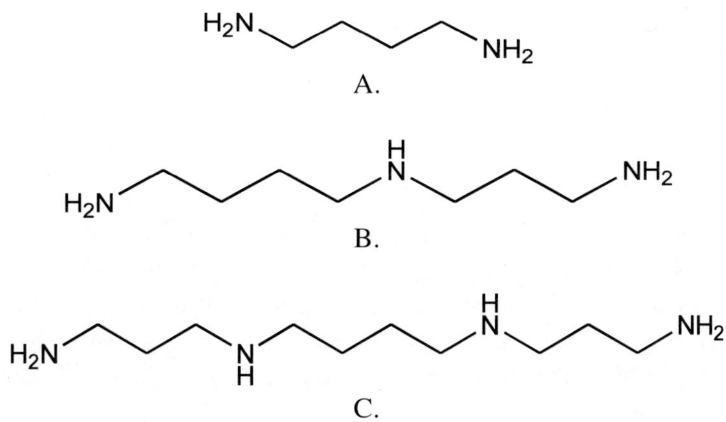

Figure 11. Structure of some common polyamines in cyanobacteria. Putrescine (A), Spermidine (B), and Spermine (C).

5.2.4. Carotenoids and Other Photoprotective Compounds

The pigment carotenoids also play an important role in photoprotection against UV-A radiation in cyanobacteria (Buckley et al. 1976, Paerl 1984, Rastogi et al. 2010c). Induction of the synthesis of some carotenoids under UV radiation strongly supports their UV photoprotective function in several cyanobacteria (Wachi et al. 1995, Ehling-Schulz et al.

1997, Ehling-Schulz and Scherer 1999, Jiang and Qiu 2005). The antioxidant function of some carotenoids has also been observed in *Trichodesmium* sp. (Kelman et al. 2009). Moreover, a number of carotenoids such as canthaxanthin, echinenone, myxoxanthophyll and zeaxanthin with protective function against photooxidative damage have been reported in several cyanobacteria (Jürgens and Weckesser 1985, Vincent and Quesada 1994, Ehling-Schulz et al. 1997, Gotz et al. 1999, Kerfeld 2004, Latifi et al. 2009, Rastogi et al. 2010c). The photoactive orange carotenoid protein (OCP) has also been found to play an important role in photoprotection under high-light conditions. It has been shown that most cyanobacteria can dissipate excess energy as heat from phycobilisomes by means of photoactive soluble OCP (Kirilovsky and Kerfeld 2012, Sedouda et al. 2014).

Besides MAAs, scytonemin and carotenoids, several other compounds such as biopterin glucoside (UVλ_{max}: 256 and 362 nm) (Matsunaga et al. 1993) and pteridines have been found in certain cyanobacteria as potent photoprotective compounds. Some other groups of compounds such as the polycationic molecules, polyamines (PAs) have also been shown to act as free radical scavengers in cyanobacteria. The diamine putrescine, triamine spermidine and tetramine spermine (Figure 11) are the most common polyamines existing inside the cells; however, spermidine represents the major polyamine in cyanobacteria (Jantaro et al. 2003, Incharoensakdi et al. 2010). The PA spermine acts as free radical scavenger and has been shown to afford the protection of ROS-induced DNA damage (Ha et al. 1998). A study in *Synechocystis* PCC 6803 showed that UV-induced cell damage was due to a decrease in spermidine content of *Synechocystis* cells (Jantaro et al. 2011). Recently, it was shown that the exogenous spermidine can alleviate UV-induced growth inhibition of *Synechocystis* PCC 6803 via reduction of H_2O_2 and malonaldehyde levels (Jantaro et al. 2014). Moreover, increased activity of PA synthesis under UV-B radiation as well as growth tolerance to UV-B stress due to increased PA contents suggest their indispensable role in mitigating the UV-induced oxidative stress in some organisms including cyanobacteria (Schweikert et al. 2011, Pothipongsa et al. 2012, Schweikert et al. 2014).

5.3. Genomic Defense Mechanisms

As discussed above, DNA is one of the prime target of UV radiation, and the restoration of damaged DNA is of utmost importance for proper functioning and existence of all life forms. A number of lesion specific repair mechanisms such as photoreactivation, excision repair and recombinational repair have been reported in several organisms (Essers et al. 2006, Rastogi et al. 2010a). In the subsequent sections, a precise mechanism of various DNA repair mechanisms is discussed.

5.3.1. Photoreativation

DNA restoration by means of photoreactivation is mediated by the important enzyme DNA photolyase using the energy of visible/blue-light. DNA photolyase enzymes such as CPD photolyase and 6-4 photolyase play an important role in removing the most cytotoxic and mutagenic DNA lesions such as CPDs or 6-4PPs, respectively. The photolyase enzyme binds precisely to the CPDs (for CPD photolyase) or 6-4PPs (for 6-4 photolyase) and directly monomerizes the cyclobutane ring of the purine/pyrimidine DNA lesion and protects the genome from damaging effects of UV radiation (Kim et al. 1992, Essen and Klar 2006). The enzyme DNA photolyases (420–616 amino acids) are monomeric flavin-dependent repair enzymes having the molecular weight of about 45-66 kDa. It consists of two known catalytic cofactors i.e. either 5,10-methenyltetrahydrofolate (MTHF) (Johnson et al. 1988), or 8-hydroxy-5-deaza-riboflavin (8-HDF) (Eker et al. 1990), and FMN (Ueda et al. 2005), and a light-harvesting cofactor such as deprotonated reduced flavin adenine dinucleotide ($FADH^-$). The cofactor MTHF or 8-HDF absorbs the long wavelength UV-A/blue light energy which is transferred to the $FADH^-$. The excited state of flavin transfers an electron to the CPD leading to a split of cyclobutane ring with transfer of electron back to the flavin (Figure 12). It has been shown that absorption of every blue light photon led to the splitting of one CPD dimer (Britt 1996). The process of UV-damaged DNA repair by means of photoreactivation has been reported in some species/strains of cyanobacteria such as *Agmenellum quadruplicatum* (Van Baalen 1968), *Plectonema boryanum* (Werbin and Rupert 1968, Saito and Werbin 1970), *Anacystis nidulans* (Asato 1972, Tang and Asato 1978), *Gloeocapsa alpicola* (*Synechocystis* PCC 6308) (O'Brien and Houghton 1982a, 1982b), *Anabaena* spp.(Levine and Thiel 1987) and *Synechocystis* PCC 6803 (Hitomi et al. 2000, Ng et al. 2000, Ng and Pakrasi 2001). Recently, Rastogi et al. (2011) have reported the photoreactivation of UV-induced T<>TCPD in *Anabaena* PCC 7937. Moreover, the photoreactivating enzymes can eliminate the majority of CPD lesion and allow the cell to perform their normal metabolic activity.

5.3.2. Excision Repair

The excision repair is a complex pathway of dark repair system that can be categorized into base excision repair (BER) and nucleotide excision repair (NER). DNA having one nucleotide lesion is removed by short-patch BER (SP-BER) while two or more nucleotide lesions are repaired by long-patch BER (LP-BER) pathway (Fortini and Dogliotti 2007). The enzyme DNA glycosylases play a key role in BER. A number of DNA glycosylases have been reported that remove different types of modified bases by cleaving the *N*-glycosidic bond between the base and the 2-deoxyribose moiety of the nucleotide residues (Rastogi et al. 2010a). The apurinic/apyrimidinic (AP) site created by a specific DNA glycosylase is removed by an AP endonuclease and remaining gap of the strand is filled and linked by DNA polymerase and DNA ligase, respectively (Seeberg et al. 1995).

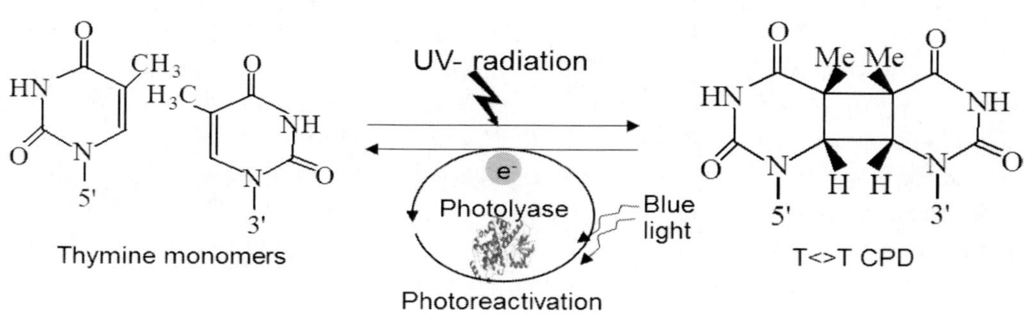

Figure 12. A simplified view of the photoreactivation mechanism (details in text).

Contrary to BER, NER is highly conserved in eukaryotes and maintains the integrity of DNA by repairing a wide range of DNA lesions, including CPDs and 6-4PPs caused by UV radiation, and some other forms of oxidative damage. Repair of DNA lesions over the entire genome is referred to as global genome NER repair (GG-NER), while repair of transcription-blocking lesions present in the transcribed DNA strands is referred to as transcription coupled NER repair (TC-NER). Similar to BER, NER also operates in a sequence i.e., damage recognition, DNA opening at damage site, dual incisions on both sides of the DNA lesion, resynthesis and ligation (Nouspikel 2009). In a prokaryotic system such as *E. coli*, six different proteins such as UvrA, B, and C (ABC-complex), UvrD (helicase II), DNA polymerase I (pol. I), and DNA ligase have been found to be involved in the NER (Lin and Sancar 1990, Orren et al. 1992). Moreover, some direct and indirect pieces of evidence have been provided demonstrating the occurrence of excision repair in some cyanobacteria (Williams et al. 1979, Geoghegan and Houghton 1982, Srivastava et al. 1971, Rastogi and Sinha 2011).

5.3.3. Recombinational Repair

The DNA double/single strand break caused by UV and other stressors is repaired by a series of complex biochemical reactions by way of recombination repair. The harmful effects on double strand breaks (DSBs) can be mitigated by two independent pathways such as homologous recombination (HR) and nonhomologous end joining (NHEJ) (Rastogi et al. 2010a). The HR requires an extensive region of sequence homology between the damaged and template DNA strands and is an error free pathway. On the other hand, NHEJ is another pathway that essentially joins broken chromosomal ends independent of sequence homology and is an error prone pathway.

Besides above mentioned repair mechanisms, some other repair systems such as SOS response and programmed cell death (PCD) have also been observed in cyanobacteria. In case of large genomic damage or failure of all specific repair systems, a SOS repair system is started by interaction of two crucial proteins such as the RecA and the LexA repressor which controls the expression of SOS genes by binding to their promoters (Courcelle et al. 2001). The occurrence of SOS system together with its complete mechanisms remain obscure in cyanobacteria; however, it has been assumed that LexA appears to function as the transcriptional regulator of the key SOS response genes in most cyanobacterial genomes (Li et al. 2010).

In case of cell damage beyond the execution of effective repair, organisms can save their life by selective cell death, also known as programmed cell death (PCD). Protection by means of PCD has been observed in several organisms including cyanobacteria such as *Trichodesmium* sp., *Microcystis aeruginosa* and *Anabaena flos-aquae* (Ning et al. 2002, Berman-Frank et al. 2004, Ross et al. 2006). In a nitrogen-fixing cyanobacterium *Trichodesmium* sp. a PCD was observed under high irradiance, iron starvation and oxidative stress (Berman-Frank et al. 2004). Moreover, the exact mechanism of UV-induced PCD in cyanobacteria is yet not very clear; however, a number of genes encoding caspases (involved in PCD) have been reported in the sequenced genome of *M. aeruginosa* (Frangeul et al. 2008). For more detailed information regarding the molecular mechanisms of UV-induced DNA damage and repair, readers are suggested to see the excellent review paper recently by Rastogi et al. (2010a).

CONCLUSION

Overall, it may be concluded that UV radiation is a highly active component of the solar radiation that brings about several physiological and biochemical alterations directly or indirectly and influences the normal life processes of cyanobacteria. However, under this unfavourable condition these photosynthetic prokaryotes have developed a number of defense mechanisms to overcome the harmful doses of UV radiation in their natural habitats.

ACKNOWLEDGMENTS

Rajesh P Rastogi is thankful to the University Grant Commission (UGC), New Delhi, India for providing Dr. D. S. Kothari Post-Doctoral Research Grant. Aran Incharoensakdi thanks Thailand National Research University Project, Office of Higher Education Commission (WCU-013-FW-57) for the support on polyamines research.

REFERENCES

Aikawa, S; Joseph, A; Yamada, R; Izumi, Y; Yamagishi, T; Matsuda, F; Kawai, H; Chang, J-S; Hasunuma, T; Kondo, A. Direct conversion of Spirulina to ethanol without pretreatment or enzymatic hydrolysis processes. *Energy Environ. Sci.* 2013, 6, 1844-1849.

Asato, Y. Isolation and characterization of ultraviolet light-sensitive mutants of the blue-green alga *Anacystis nidulans*. *J. Bacteriol.* 1972, 110, 1058-1064.

Badger, MR; von Caemmerer, S; Ruuska, S; Nakano H. Electron flow to oxygen in higher plants and algae: rates and control of direct photoreduction (Mehler reaction) and rubsico oxygenase. *Philos. Trans R Soc. Lond B; Biol. Sci.* 2000, 355, 1433-1446.

Balskus, EP; Walsh, CT. Investigating the initial steps in the biosynthesis of cyanobacterial sunscreen scytonemin. *J. Am. Chem. Soc.* 2008, 130, 15260-15261.

Balskus, EP; Walsh, CT. An enzymatic cyclopentyl[b]indole formation involved in scytonemin biosynthesis. *J. Am. Chem. Soc.* 2009, 131, 14648-14649.

Balskus, EP; Walsh, CT. The genetic and molecular basis for sunscreen biosynthesis in cyanobacteria. *Science* 2010, 329, 1653-1656.

Balskus, EP; Case, RJ; Walsh, CT. The biosynthesis of cyanobacterial sunscreen scytonemin in intertidal microbial mat communities. *FEMS Microbiol. Ecol.* 2011, 77, 322-332.

Banerjee, M; Häder, DP. Effects of UV radiation on the rice field cyanobacterium, *Aulosira ferlxissima*. *Environ. Exp. Bot.* 1996, 36, 281-291.

Bebout, BM; Garcia-Pichel, F. UV B-induced vertical migrations of cyanobacteria in a microbial mat. *Appl. Environ. Microbiol.* 1995, 61, 4215-4222.

Berman-Frank, I; Bidle, KD; Haramaty, L; Falkowski, PG. The demise of the cyanobacterium, *Trichodesmium* spp., via an autocatalyzed cell death pathway. *Limnol. Oceanogr.* 2004, 49, 997-1005.

Bernroitner, M; Zamocky, M; Furtmüller, PG; Peschek, GA; Obinger, C. Occurrence, phylogeny, structure, and function of catalases and peroxidases in cyanobacteria. *J. Exp. Bot.* 2009, 60, 423-440.

Bhandari, R; Sharma, PK. High-light-induced changes on photosynthesis, pigments, sugars, lipids and antioxidant enzymes in freshwater (*Nostoc spongiaeforme*) and marine (*Phormidium corium*) cyanobacteria. *Photochem. Photobiol.* 2006, 82, 702-710.

Bhandari, RR; Sharma, PK. Photosynthetic and biochemical characterization of pigments and UV-absorbing compounds in *Phormidium tenue* due to UV-B radiation. *J. Appl. Phycol.* 2011, 23, 283-292.

Bhargava, P; Atri, N; Srivastava, AK; Rai, LC. Cadmium mitigates ultraviolet-B stress in *Anabaena doliolum*: enzymatic and non-enzymatic antioxidants. *Biol. Plant.* 2007, 51, 546-550.

Bhaya, D; Bianco, NR; Bryant, D; Grossman, A. Type IV pilus biogenesis and motility in the cyanobacterium *Synechocystis* sp. PCC6803. *Mol. Microbiol.* 2000, 37, 941-951.

Bhaya, D. Light matters: phototaxis and signal transduction in unicellular cyanobacteria. *Mol. Microbiol.* 2004, 53, 745-754.

Blakefield, MK; Harris, DO. Delay of cell differentiation in *Anabaena aequalis* caused by UV-B radiation and the role of photoreactivation and excision repair. *Photochem. Photobiol.* 59: 1994, 204-208.

Britt, AB. DNA damage and repair in plants. *Annu. Rev. Plant Physiol. Plant Mol. Biol.* 1996, 47, 75-100.

Brocks, JJ; Logan, GA; Buick, R; Summons, RE. Archean molecular fossils and the early rise of eukaryotes. *Science* 1999, 285, 1033-1036.

Buckley, CE; Houghton, JA. A study of the effects of near UV radiation on the pigmentation of the blue-green alga *Gloeocapsa alpicola*. *Arch. Microbiol.* 1976, 107, 93-97.

Büdel, B. Ecology and diversity of rock-inhabiting cyanobacteria in tropical regions. *Eur. J. Phycol.* 1999, 34, 361-370.

Bultel-Poncé, V; Felix-Theodore, F; Sarthon, C; Ponge, J-F; Bodo, B. New pigments from the terrestrial cyanobacterium *Scytonema* sp. collected on the Mitaraka Inselberg, French Guyana. *J. Nat. Prod.* 2004, 67, 678-681.

Cameron, JC; Pakrasi, HB. Essential role of glutathione in acclimation to environmental and redox perturbations in the cyanobacterium *Synechocystis* sp. PCC 68031[W][OA]. *Plant Physiol.* 2010, 154, 1672-1685.

Campbell, D; Eriksson, MJ, Öquist, G; Gustafsson, P; Clarke, AK. The cyanobacterium *Synechococcus* resists UV-B by exchanging photosystem II reaction-center D1 proteins. *Proc. Natl. Acad. Sci. USA* 1998, 95, 364-369.

Carpenter, EJ; Romans, K. Major role of the cyanobacterium *Trichodesmium* in nutrient cycling in the North Atlantic Ocean. *Science* 1991, 254, 1356-1358.

Castenholz, RW. Multiple strategies for UV tolerance in cyanobacteria. *Spectrum* 1997, 10, 10-16.

Castenholz, RW. Motility and taxes. In: Carr NG, Whitton BA (eds). The Biology of Cyanobacteria. Berkeley: University of California Press, pp. 413-440.

Chen, LZ; Wang, GH; Hong, S; Liu, A; Li, C; Liu, YD. UV-B-induced oxidative damage and protective role of exopolysaccharides in desert cyanobacterium *Microcoleus vaginatus*. *J. Integr. Plant Biol.* 2009, 51, 194-200.

Chris, A; Zeeshan, M; Abraham, G; Prasad, SM. Proline accumulation in *Cylindrospermum* sp. *Environ. Exp. Bot.* 2006, 57, 154-159.

Cockell, CS; Rettberg, P; Rabbow, E; Olsson-Francis, K. Exposure of phototrophs to 548 days in low Earth orbit: Microbial selection pressures in outer space and on early earth. *ISME J.* 2011, 5, 1671-1682.

Conde, FR; Churio, MS; Previtali, CM. The photoprotector mechanism of mycosporine-like amino acids. Excited-state properties and photostability of porphyra-334 in aqueous solution. *J. Photochem. Photobiol.* B 2000, 56, 139-144.

Conde, FR; Churio, MS; Previtali, CM. The deactivation pathways of the excited-states of the mycosporine-like amino acids shinorine and porphyra-334 in aqueous solution. *Photochem. Photobiol. Sci.* 2004, 3, 960-967.

Courcelle, J, Khodursky, A, Peter, B; Brown, PO; Hanawalt, PC. Comparative gene expression profiles following UV exposure in wild-type and SOS-deficient *Escherichia coli*. *Genetics*, 2001, 58, 41-64.

De Caiola, MG; Billi, D; Friedmann, EI. Effect of desiccation on envelopes of the cyanobacterium *Chroococcidiopsis* sp. (Chroococcales). *Eur. J. Phycol.* 1996, 31, 97-105.

de la Coba, F; Aguilera, J; Figueroa, FL; de Gálvez, MV; Herrera, E. Antioxidant activity of mycosporine-like amino acids isolated from three red macroalgae and one marine lichen. *J. Appl. Phycol.* 2009, 21, 161-169.

Dexter, J; Fu, P. Metabolic engineering of cyanobacteria for ethanol production. *Energy Environ. Sci.* 2009, 2, 857-864.

Dillon, JG; Tatsumi, CM; Tandingan, PG; Castenholz, RW. Effect of environmental factors on the synthesis of scytonemin, a UV-screening pigment, in a cyanobacterium (*Chroococcidiopsis* sp.). *Arch. Microbiol.* 2002, 177, 322-331.

Doetsch, PW; Zastawny, TH; Martin, AM; Dizdaroglu, M. Monomeric base damage products from adenine, guanine and thymine induced by exposure of DNA to ultraviolet radiation. *Biochemistry* 1995, 34, 737-742.

Donkor, V; Häder, D-P. Effects of solar and ultraviolet radiation on motility, photomovement and pigmentation in filamentous, gliding cyanobacteria. *FEMS.* 1991, 86, 2159-2168.

Donkor, VA; Amewowor, DHAK; Häder, D-P. Effects of tropical solar radiation on the motility of filamentous cyanobacteria. *FEMS Microbiol. Ecol.* 1993, 12, 143-148.

Dunlap, WC; Yamamoto, Y. Small-molecule antioxidants in marine organisms: antioxidant activity of mycosporine-glycine. *Comp. Biochem. Physiol. B Biochem. Mol. Biol.* 1995, 112, 105-114.

Dutta, D; De, D; Chaudhuri, S; Bhattacharaya, SK. Hydrogen production by cyanobacteria. *Microb. Cell Fact.* 2005, 4, 1-11.

Ehling-Schulz, M; Bilger, W; Scherer, S. UV-B induced synthesis of photoprotective pigments and extracellular polysaccharides in the terrestrial cyanobacterium *Nostoc commune*. *J. Bacteriol.* 1997, 179, 1940-1945.

Ehling-Schulz, M; Scherer, S. UV protection in cyanobacteria. *Eur. J. Phycol.* 1999, 34, 329-338.

Ehling-Schulz, M; Schulz, S; Wait, R; Görg, A; Scherer, S. The UV-B stimulon of the terrestrial cyanobacterium *Nostoc commune* comprises early shock proteins and late acclimation proteins. *Mol. Microbiol.* 2002, 46, 827-843.

Eker, APM; Kooiman, P; Hessels, JKC; Yasui, A. DNA photoreactivating enzyme from the cyanobacterium *Anacystis nidulans*. *J. Biol. Chem.* 1990, 265, 8009-8015.

Essen, LO; Klar, T. Light-driven DNA repair by photolyases. *Cell. Mol. Life Sci.* 2006, 63, 1266-1277.

Essers, J; Vermeulen, W; Houtsmuller, AB. DNA damage repair: anytime, anywhere? *Curr. Opin. Cell Biol.* 2006, 18, 240- 246.

Fischer, WF. Life before the rise of oxygen. *Nature* 2008, 455, 1051-1052.

Fleming, ED; Castenholz, RW. Effects of periodic desiccation on the synthesis of the UV screening compound, scytonemin, in cyanobacteria. *Environ. Microbiol.* 2007, 9, 1448-1455.

Fortini, P; Dogliotti, E. Base damage and single-strand break repair: mechanisms and functional significance of short- and long-patch repair subpathways. *DNA Repair* 2007, 6, 398-409.

Frangeul, L; Quillardet, P; Castets, AM; Humbert, JF; Matthijs, HCP; Cortez, D; Tolonen, A; Zhang, CC; Gribaldo, S; Kehr, JC; Zilliges, Y; Ziemert, N; Becker, S; Talla, E; Latifi, A; Billault, A; Lepelletier, A; Dittmann, E; Bouchier, C; Tandeau de Marsac, N. Highly plastic genome of *Microcystis aeruginosa* PCC 7806, a ubiquitous toxic freshwater cyanobacterium. *BMC Genomics* 2008, 9, 274.

Fridovich, I. Superoxide anion radiacal (O_2^-), superoxide dismutases and related matters. *J. Biol. Chem.* 1997, 272, 18515-18517.

Gaber, A; Yoshimura, K; Tamoi, M; Takeda, T; Nakano, Y; Shigeoka, S. Induction and functional analysis of two reduced nicotinamide adenine dinucleotide phosphate-dependent glutathione peroxidase-like proteins in *Synechocystis* PCC 6803 during the progression of oxidative stress. *Plant Physiol.* 2004, 136, 2855-2861.

Gantar, M; Svircev, Z. Microalgae and cyanobacteria: food for thought. *J. Phycol.* 2008, 44, 260-268.

Gao, K; Yu, H; Brown, MT. Solar PAR and UV radiation affects the physiology and morphology of the cyanobacterium, *Anabaena* spp. PCC 7120. *J. Photochem. Photobiol. B* 2007, 89, 117-124.

Gao, K; Li, P; Watanabe, T; Helbling, EW. Combined effects of ultraviolet radiation and temperature on morphology, photosynthesis, and DNA of *Arthrospira* (*Spirulina*) *platensis* (Cyanophyta). J. Phycol. 2008, 44, 777-786.

Garcia-Pichel, F; Castenholz, RW. Characterization and biological implications of scytonemin, a cyanobacterial sheath pigment. *J. Phycol.* 1991, 27, 395-409.

Garcia-Pichel, F; Sherry, ND; Castenholz, RW. Evidence for an ultraviolet sunscreen role of the extracellular pigment scytonemin in the terrestrial cyanobacterium *Chlorogloeopsis* sp., *Photochem. Photobiol.* 1992, 56, 17-23.

Garcia-Pichel, F; Wingard, CE; Castenholz, RW. Evidence regarding the UV sunscreen role of a mycosporine-like compound in the cyanobacterium *Gloeocapsa* sp. *Appl. Environ. Microbiol.* 1993, 59, 170-176.

Ge, H; Xia, L; Zhou, X; Zhang, D; Hu, C. Effects of light intensity on components and topographical structures of extracellular polysaccharides from the cyanobacteria *Nostoc* sp. *J. Microbiol.* 2014, 52, 179-183.

Geoghegan, C; Houghton, JA. Isolation of a UV endonuclease from the cyanobacterium *Synechocystis* PCC 6308. *J. Gen. Microbiol.* 1982, 128, 1743-1747.

Giordanino, MVF, Strauch, SM, Villafane, VE, Helbling, EW. Influence of temperature and UVR on photosynthesis and morphology of four species of cyanobacteria. *J. Photochem. Photobiol. B* 2011, 103, 68-77.

Gotz, T., Windhovel, U., Boger, P., Sandmann, G. Protection of photosynthesis against UV-B radiation by carotenoids in transformants of the cyanobacterium *Synechococcus* PCC 7942. *Plant Physiol.* 1999, 120, 599-604.

Grant, CS; Louda, JW. Scytonemin-imine, a mahogany-colored UV/Vis sunscreen of cyanobacteria exposed to intense solar radiation. *Organic Geochem.* 2013, 65, 29-36.

Gröniger, A; Häder, D-P. Stability of mycosporine-like amino acids. *Recent Res. Dev. Photochem. Photobiol.* 2000, 4, 247-252.

Gupta, R; Bhadauriya, P; Chauhan, VS; Bisen, PS. Impact of UV-B radiation on thylakoid membrane and fatty acid profile of *Spirulina platensis*. *Curr. Microbiol.* 2008, 56, 156-161.

Ha, HC; Sirisoma, NS; Kuppusamy, P; Zweir, JL; Woster, PM; Casero Jr RA. The natural polyamine spermine functions directly as a free radical scavenger. *Proc. Natl. Acad. Sci. USA* 1998, 95, 11140-11145.

Häder, D-P. Effects of UV-B on motility and photoorientation in the cyanobacterium, *Phormidium uncinatum*. *Arch. Microbiol.* 1984, 140, 34- 39.

Häder, D-P; Worrest RC. Effects of enhanced solar radiation on aquatic ecosystems. *Photochem. Photobiol.* 1991, 53, 717-725.

Häder, D-P; Helbling, EW; Williamson, CE; Worrest, RC. Effects of UV radiation on aquatic ecosystems and interactions with climate change. *Photochem. Photobiol. Sci.* 2011, 10, 242-260.

Hall, DB; Holmlin, RE; Barton, JK. Oxidative DNA damage through long range electron transfer. *Nature* 1996, 382, 731-735.

Halliwell, B; Gutteridge, J. Free radicals in biology and medicine, Oxford University Press, Oxford, UK, 4th edition, 2007.

Han, PP; Sun, Y; Jia, S; Zhong, C; Tan, Z-L. Effects of light wavelengths on extracellular and capsular polysaccharide production by *Nostoc flagelliforme*. *Carbohydrate Polymers* 2014, 105, 145-151.

He, Y-Y; Häder, D-P. UV-B-induced formation of reactive oxygen species and oxidative damage of the cyanobacterium *Anabaena* sp.: protective effects of ascorbic acid and N-acetyl-L-cysteine. *J. Photochem. Photobiol. B* 2002a, 66, 115-124.

He, Y-Y; Häder, D-P. Reactive oxygen species and UV-B: effect on cyanobacteria. *Photochem. Photobiol. Sci.* 2002b, 1, 729-736.

Hitomi, K; Okamoto, K; Daiyasu, H; Miyashita, H; Iwai, S; Toh, H; Ishiura, M; Todo, T. Bacterial cryptochrome and photolyase: characterization of two photolyase-like genes of *Synechocystis* sp. PCC6803. *Nucleic Acids Res.* 2000, 28, 2353-2362.

Incharoensakdi, A; Jantaro, S; Raksajit, W; Maenpaa P. Polyamines in cyanobacteria: biosynthesis, transport and abiotic stress response. In: Mendez-Vilas A (ed.). Current Research, Technology and Education. Topics in applied microbiology and microbial miotechnology, Microbiology Book series No. 2, Vol. 1, Formatex, Spain, 2010, pp. 23-32.

Ishikawa, T; Shigeoka, S. Recent advances in ascorbate biosynthesis and the physiological significance of ascorbate peroxidase in photosynthesizing organisms. *Biosci. Biotechnol. Biochem.* 2008, 72, 1143-1154.

Ismaiel, MMS; El-Ayouty, YM; Loewen, PC; Piercey-Normore, MD. Characterization of the iron-containing superoxide dismutase and its response to stress in cyanobacterium *Spirulina* (*Arthrospira*) *platensis*. *J. Appl. Phycol.* 2014, DOI 10.1007/s10811-013-0233-y.

Iturriaga, R; Mitchell, BG. Chroococcoid cyanobacteria: a significant component in the food web dynamics of the open ocean. *Mar. Ecol. Prog. Ser.* 1986, 28, 291-297.

Jantaro, S; Maenpaa, P; Mulo, P; Incharoensakdi, A. Content and biosynthesis of polyamines in salt and osmotically stressed cells of *Synechocystis* sp. PCC 6803. *FEMS Microbiol. Lett.* 2003, 228, 129-135.

Jantaro, S; Pothipongsa, A; Khanthasuwan, S; Incharoensakdi, A. Short-term UV-B and UV-C radiations preferentially decrease spermidine contents and arginine decarboxylase transcript levels of *Synechocystis* sp. PCC 6803. *Curr. Microbiol.* 2011, 62, 420-426.

Jantaro, S; Baebprasert, W; Piyamawadee, C; Sodsuay, O; Incharoensakdi, A. Exogenous spermidine alleviates UV-induced growth inhibition of *Synechocystis* sp. PCC 6803 via reduction of hydrogen peroxide and malonaldehyde levels. *Appl. Biochem. Biotechnol.* 2014, DOI 10.1007/s12010-014-0887-1.

Jiang, H; Qiu, B. Photosynthetic adaptation of a bloom-forming cyanobacterium *Microcystis aeruginosa* (cyanophyceae) to prolonged UV-B exposure. *J. Phycol.* 2005, 41, 983-992.

Johnson, JL; Hamm-Alvarez, S; Payne, G; Sancar, GB; Rajagopalan, KV; Sancar, A. Identification of the second chromophore of *Escherichia coli* and yeast DNA photolyases as 5,10-methenyltetrahydrofolate. *Proc. Natl. Acad. Sci. USA* 1988, 85, 2046-2050.

Jürgens, UJ; Weckesser, J. Carotenoid-containing outer membrane of *Synechocystis* sp. strain PCC 6714. *J. Bacteriol.* 1985, 164, 384-389.

Karni, L; Moss, SJ; Tel-Or, E. Glutathione reductase activity in heterocysts and vegetative cells of *Nostoc muscorum*. *Arch. Microbiol.* 1984, 140, 215-217.

Karsten, U; Lembcke, S; Schumann, R. The effects of ultraviolet radiation on photosynthetic performance, growth and sunscreen compounds in aeroterrestrial biofilm algae isolated from building facades. *Planta* 2007, 225, 991-1000.

Kelman, D; Ben-Amotz, A; Berman-Frank, I. Carotenoids provide the major antioxidant defence in the globally significant N_2-fixing marine cyanobacterium *Trichodesmium*. *Environ. Microbiol.* 2009, 11, 1897-1908.

Ke, W-T; Dai, G-Z; Jiang, H-B; Zhang, R; Qiu, B-S. Essential roles of iron superoxide dismutase in photoautotrophic growth of *Synechocystis* sp. PCC 6803 and heterogeneous expression of marine *Synechococcus* sp. CC9311 copper/zinc superoxide dismutase within its *sodB* knockdown mutant. *Microbiology* 2014, 160, 228-241.

Kerfeld, AC. Water-soluble carotenoid proteins of cyanobacteria. *Arch. Biochem. Biophys.* 2004, 430, 2-9.

Khetkorn, W; Khanna, N; Incharoensakdi, A; Lindblad, P. Metabolic and genetic engineering of cyanobacteria for enhanced hydrogen production. *Biofuels* 2013, 4, 535-561.

Kim, S-T; Heelis, PF; Sancar, A. Energy transfer (deazaflavin → FADH2) and electron transfer (FADH2→T<>T) kinetics in *Anacystis nidulans* photolyase. *Biochemistry* 1992, 31, 11244-11248.

Kirilovsky, D; Kerfeld, CA. The orange carotenoid protein in photoprotection of photosystem II in cyanobacteria. *Biochim. Biophysic Acta* 2012, 1817, 158-166.

Kolowrat, C; Partensky, F; Mella-Flores, D; Le Corguillé, G; Boutte, C; Blot, N; Ratin, M; Ferréo, M; Lecomte, X; Gourvil, P; Lennon, J-F; Kehoe, DM; Garczarek, L. Ultraviolet stress delays chromosome replication in light/dark synchronized cells of the marine cyanobacterium *Prochlorococcus marinus* PCC9511. *BMC Microbiol.* 2010, 10, 204.

Kruschel, C; Castenholz, RW. The effect of solar UV and visible irradiance on the vertical movements of cyanobacteria in microbial mats of hypersaline waters. *FEMS Microbiol.* Ecol. 1998, 27, 53-72.

Kumar, A; Tyagi, MB; Jha PN. Evidences showing ultraviolet-B radiation-induced damage of DNA in cyanobacteria and its detection by PCR assay. *Biochem. Biophys. Res. Commun.* 2004, 318, 1025-1030.

Kumar, A; Tyagi, MB; Singh, N; Tyagi, R; Jha, PN; Sinha, RP; Häder, D-P. Role of white light in reversing UV-B-mediated effects in the N_2-fixing cyanobacterium *Anabaena* BT2. *J. Photochem. Photobiol.* B 2003,71, 35-42.

Kumar, A; Sinha, RP; Häder, D-P. Effect of UV-B on enzymes of nitrogen metabolism in the cyanobacterium *Nostoc calcicola*. *J. Plant Physiol.* 1996, 148, 86-91.

Latifi, A; Ruiz, M; Zhang, CC. Oxidative stress in cyanobacteria. *FEMS Microbiol. Rev.* 2009, 33, 258-278.

Lesser, MP; Stochaj, WR. Photoadaptation and protection against active forms of oxygen in the symbiotic prokaryote *prochloron* sp. and its ascidian host. *Appl. Environ. Microbiol.* 1990, 56, 1530-1535.

Lesser, MP. Effects of ultraviolet radiation on productivity and nitrogen fixation in the cyanobacterium, *Anabaena* sp. (Newton's strain). *Hydrobiologia* 2008, 598, 1-9.

Levine, E; Thiel, T. UV-Inducible DNA repair in the cyanobacteria *Anabaena* spp. *J. Bacteriol.* 1987, 169, 3988-3993.

Li, S; Xu, M; Su, Z. Computational analysis of LexA regulons in cyanobacteria. *BMC Genomics* 2010, 11, 527.

Li, H; Xu, J; Liu, Y; Ai, S; Qin, F; Li, Z; Zhang, H; Huang, Z. Antioxidant and moisture-retention activities of the polysaccharide from *Nostoc commune*. *Carbohydrate Polymers* 2011, 83, 1821-1827.

Lin, J-J; Sancar, A. Reconstitution of nucleotide excision nuclease with UvrA and UvrB proteins from *Escherichia coli* and UvrC protein from *Bacillus subtilis*. *J. Biol. Chem.* 1990, 265, 21337-21341.

Ma, Z; Gao, K. Spiral breakage and photoinhibition of *Arthrospira platensis* (cyanophyta) caused by accumulation of reactive oxygen species under solar radiation. *Environ. Exp. Bot.* 2010, 68, 208-213.

Manney, GL; Santee, ML; Rex, M; Livesey, NJ; Pitts, MC; et al. Unprecedented Arctic ozone loss in 2011. *Nature* 2011, 478, 469-475.

Matsunaga, T; Burgess, JG; Yamada, N; Komatsu, K; Yoshida, S; Wachi, Y. An ultraviolet (UV-A) absorbing biopterin glucoside from the marine planktonic cyanobacterium *Oscillatoria* sp. *Appl. Microbiol. Biotechnol.* 1993, 39, 250-253.

McKenzie, RL; Aucamp, PJ; Bais, AF; Björn, LO; Ilyase, M. Changes in biologically-active ultraviolet radiation reaching the earth's surface. *Photochem. Photobiol. Sci.* 2007, 6, 212-217.

Mehler, AH. Studies on reactions of illuminated chloroplasts. I. Mechanisms of the reduction of oxygen and other hill reagents. *Arch. Biochem. Biophys.* 1951, 33, 65-77.

Misonou, T; Saitoh, J; Oshiba, S; Tokitomo, Y; Maegawa, M; Inoue, Y; Hori, H; Sakurai, T. UV-absorbing substance in the red alga *Porphyra yezoensis* (Bangiales, Rhodophyta) block thymine photodimer production. *Mar. Biotechnol.* 2003, 5, 194-200.

Miyake, C; Michihata, F; Asada, K. Scavenging of hydrogen peroxide in prokaryotic and eukaryotic algae. *Plant Cell Physiol*. 1991, 32, 33-43.

Mota, R; Guimaraes, R; Büttel, Z; Rossi, F; Colica, G; Silva, CJ; Santos, C; Gales, L; Zille, A; De Philippis, R; Pereira, SB; Tamagnini, P. Production and characterization of extracellular carbohydrate polymer from *Cyanothece* sp. CCY 0110. *Carbohydrate Polymers* 2013, 92, 1408-1415.

Mutsuda, M; Ishikawa, T; Takeda, T; Shigeoka S. The catalase–peroxidase of *Synechococcus* PCC 7942: purification, nucleotide sequence analysis and expression in *Escherichia coli*. *Biochem. J.* 1996, 316, 251-257.

Nadeau, T-L; Howard-williams, C; Castenholz, RW. Effects of solar UV and visible irradiance on photosynthesis and vertical migration of *Oscillatoria* sp. (Cyanobacteria) in an Antarctic microbial mat. *Aquatic Microbial Ecol*. 1999, 20, 231-243.

Ng, W-O; Zentella, R; Wang, Y; Taylor, JSA; Pakrasi, HB. phrA, the major photoreactivating factor in the cyanobacterium *Synechocystis* sp. strain PCC 6803 codes for a cyclobutane–pyrimidine-dimer-specific DNA photolyase. *Arch. Microbiol*. 2000, 173, 412-417.

Ng, W-O; Pakrasi, HB. DNA photolyase homologs are the major UV resistance factors in the cyanobacterium *Synechocystis* sp. PCC 6803. *Mol. Gen. Genet*. 2001, 264, 924-930.

Nicolaus, B; Panico, A; Lama, L; Romano, I; Manca MC; Giulio, AD; Gambacorta, A. Chemical composition and production of exopolysaccharides from representative members of heterocystous and non-heterocystous cyanobacteria. *Phytochemistry* 1999, 52, 639-647.

Ning, SB; Guo, HL; Wang, L; Song, YC. Salt stress induces programmed cell death in prokaryotic organism *Anabaena*. *J. Appl. Microbiol*. 2002, 93, 15-28.

Nishiyama, Y; Allakhverdiev, SI; Murata, N. A new paradigm for the action of reactive oxygen species in the photoinhibition of photosystem II. *Biochim. Biophys. Acta – Bioenergetics* 2006, 1757, 742-749.

Nouspikel, T. Nucleotide excision repair: variations on versatility. *Cell Mol. Life Sci*. 2009, 66, 994-1009.

O'Brien, PA; Houghton, JA. UV induced DNA degradation in the cyanobacterium *Synechocystis* PCC 6308. *Photochem. Photobiol*. 1982a, 36, 417-422.

O'Brien, PA; Houghton JA. Photoreactivation and excision repair on UV induced pyrimidine dimers in the unicellular cyanobacterium *Gloeocapsa alpicola* (*Synechocystis* PCC 6308). *Photochem. Photobiol*. 1982b, 35, 359-364.

Olson, JM. Photosynthesis in the Archean era. *Photosyn. Res*. 2006, 88, 109-117.

Olsson-Francis, K; Watson, JS; Cockell CS. Cyanobacteria isolated from the high-intertidal zone: a model for studying the physiological prerequisites for survival in low Earth orbit. *Inter. J. Astrobiol*. 2013, 12, 292-303.

Orren, DK; Selby, P; Hearst, JE; Sancar A. Post incision steps of nucleotide excision repair in *Escherichia coli*. Disassembly of the UvrBC-DNA complex by helicase II and DNA polymerase I. *J. Biol. Chem*. 1992, 267, 780-788.

Paerl, HW. Cyanobacterial carotenoids: their roles in maintaining optimal photosynthetic production among aquatic bloom forming genera. *Oecologia* 1984, 61, 143-149.

Parikh, A; Madamwar, D. Partial characterization of extracellular polysaccharides from cyanobacteria. *Bioresour. Technol*. 2006, 97, 1822-1827.

Parmar, A; Singh, NK; Pandey, A; Gnansounou, E; Madamwar, D. Cyanobacteria and microalgae: a positive prospect for biofuels. *Bioresour. Technol.* 2011, 102, 10163-10172.

Patterson, CO; Myers, J. Photosynthesis production of hydrogen peroxide by *Anacystis nidulans*. *Plant Physiol.* 1973, 51, 104-109.

Pereira, S; Zille, A; Micheletti, E; Moradas-Ferreira, P; De Philippis, R; Tamagnini, P. Complexity of cyanobacterial exopolysaccharides: composition, structures, inducing factors and putative genes involved in their biosynthesis and assembly. *FEMS Microbiol. Rev.* 2009, 33, 917-941.

Pothipongsa, A, Jantaro, S, Incharoensakdi, A. Polyamines induced by osmotic stress protect *Synechocystis* sp. PCC 6803 cells and arginine decarboxylase transcripts against UV-B radiation. *Appl. Biochem. Biotechnol.* 2012, 168, 1476-1488.

Potts, M. Minireview: mechanism of desiccation tolerance in cyanobacteria. *Eur. J. Phycol.* 1999, 34, 319-328.

Prasad, SM; Zeeshan, M. Effect of UV-B and monocrotophos, singly and in combination, on photosynthetic activity and growth of non-heterocystous cyanobacterium *Plectonema boryanum*. *Environ. Exp. Bot.* 2004, 52, 175-184.

Priya, B; Sivaprasanth, RK; Jensi, VD; Uma, L; Subramanian, G; Prabaharan, D. Characterization of manganese superoxide dismutase from a marine cyanobacterium *Leptolyngbya valderiana* BDU20041. *Saline Systems* 2010, 6, 6.

Priya, B; Premanandh, J; Dhanalakshmi, RT; Seethalakshmi, T; Uma, L; Prabaharan, D; Subramanian, G. Comparative analysis of cyanobacterial superoxide dismutases to discriminate canonical forms. *BMC Genomics* 2007, 8, 435.

Proteau, PJ; Gerwick, WH; Garcia-Pichel, F; Castenholz, R. The structure of scytonemin, an ultraviolet sunscreen pigment from the sheaths of cyanobacteria. *Experientia* 1993, 49, 825-829.

Quesada, A; Vincent, WF; Lean, DRS. Community and pigment structure of Arctic cyanobacterial assemblages: the occurrence and distribution of UV-absorbing compounds. *FEMS Microbiol. Ecol.* 1999, 28, 315-323.

Quesada, A; Vincent, WF. Strategies of adaptation by Antarctic cyanobacteria to ultraviolet radiation. *Eur. J. Phycol.* 1997, 32, 335-342.

Rai, LC; Tyagi, B; Rai, PK; Mallick, N. Interactive effects of UV-B and heavy metals (Cu and Pb) on nitrogen and phosphorus metabolism of a N_2-fixing cyanobacterium *Anabaena doliolum*. *Environ. Exp. Bot.* 1998, 39, 221-231.

Ramsing, NB; Prufert-Bebout, L. Motility of *Microcoleus chthonoplastes* subjected to different light intensities quantified by digital image analysis. In: Stal, LJ; Caumette, P. (eds.). Microbial Mats, Structure, Development and Environmental Significance. NATO ASI series, Springer-Verlag, Berlin, pp. 183-191.

Rastogi, RP; Sinha, RP. Biotechnological and industrial significance of cyanobacterial secondary metabolites. *Biotechnol. Adv.* 2009, 27, 521-539.

Rastogi, RP; Richa; Kumar, A; Tyagi, MB; Sinha, RP. Molecular mechanisms of ultraviolet radiation-induced DNA damage and repair. *J. Nucleic Acids* 2010, Article ID 592980, doi:10.4061/2010/592980, (2010a).

Rastogi, RP; Singh, SP; Häder, D-P; Sinha, RP. Detection of reactive oxygen species (ROS) by the oxidant sensing probe 2′,7′-dichlorodihydrofluorescein diacetate in the cyanobacterium *Anabaena variabilis* PCC 7937. *Biochem. Biophys. Res. Commun.* 2010b, 397, 603-607.

Rastogi, RP; Richa; Sinha, RP; Singh, SP; Häder, D-P. Photoprotective compounds from marine organisms. *J. Ind. Microbiol. Biotechnol.* 2010c, 37, 537-558.

Rastogi, RP; Richa; Singh, SP; Häder, D-P; Sinha, RP. Mycosporine-like amino acids profile and their activity under PAR and UVR in a hot-spring cyanobacterium *Scytonema* sp. HKAR-3. *Aus. J. Bot.* 2010d, 58, 286-293.

Rastogi, RP; Singh, SP; Häder, D-P; Sinha, RP. Ultraviolet-B-induced DNA damage and photorepair in the cyanobacterium *Anabaena variabilis* PCC 7937. *Environ. Exp. Bot.* 2011, 74, 280-288.

Rastogi, RP; Sinha, RP. Solar ultraviolet radiation-induced DNA damage and protection/repair strategies in cyanobacteria. *Int. J. Pharm. Biosci.* 2011, 2, 271-288.

Rastogi, RP; Kumari, S; Richa; Han, T; Sinha, RP. Molecular characterization of hot spring cyanobacteria and evaluation of their photoprotective compounds. *Can. J. Microbiol.* 2012, 58, 719-727.

Rastogi, RP; Sinha, RP; Incharoensakdi, A. Partial characterization, UV-induction and photoprotective function of sunscreen pigment, scytonemin from *Rivularia* sp. HKAR-4. *Chemosphere* 2013, 93, 1874-1878.

Rastogi, RP; Singh, SP; Incharoensakdi, A; Häder, D-P; Sinha, RP. Ultraviolet radiation-induced generation of reactive oxygen species, DNA damage and induction of UV-absorbing compounds in the cyanobacterium *Rivularia* sp. HKAR-4. *South Afr. J. Bot.* 2014, 90, 163-169.

Rastogi, RP; Incharoensakdi, A. UV radiation-induced biosynthesis, stability and antioxidant activity of mycosporine-like amino acids (MAAs) in a unicellular cyanobacterium *Gloeocapsa* sp. CU2556. *J. Photochem. Photobiol.* B 2014a, 130, 287-292.

Rastogi, RP; Incharoensakdi, A. Characterization of UV-screening compounds, mycosporine-like amino acids, and scytonemin in the cyanobacterium *Lyngbya* sp. CU2555. *FEMS Microbiol. Ecol.* 2014b, 87, 244-256.

Rastogi, RP; Incharoensakdi, A. Analysis of UV-absorbing photoprotectant mycosporine-like amino acid (MAA) in the cyanobacterium *Arthrospira* sp. CU2556. *Photochem. Photobiol. Sci.* 2014c, 13, 1016-1024.

Rath, J; Adhikari, SP. Response of the estuarine cyanobacterium *Lyngbya aestuarii* to UV-B radiation. *J. Appl. Phycol.* 2007, 19, 529-536.

Reynolds, CS; Oliver, RL; Walsby, AE. Cyanobacterial dominance: the role of buoyancy regulation in dynamic lake environments. *N. Z. J. Mar. Freshwater Res.* 1987, 21, 379-390.

Richardson, LL; Castenholz, RW. Diel vertical movements of the cyanobacterium *Oscillatoria terebriformis* in a sulfide-rich hot spring mat. *Appl. Environ. Microbiol.* 1987, 53; 2142-2150.

Rinalducci, S; Hideg, E; Vass, I; Zolla, L. Effect of moderate UV-B irradiation on *Synechocystis* PCC 6803 Biliproteins. *Biochem. Biophys. Res. Commun.* 2006, 341, 1105-1112.

Ross, C; Santiago-Vázquez, L; Paul, V. Toxin release in response to oxidative stress and programmed cell death in the cyanobacterium *Microcystis aeruginosa*. *Aquatic Toxicol.* 2006, 78, 66-73.

Saito, N; Werbin, H. Purification of a blue-green algal deoxyribonucleic acid photoreactivating enzymne. An enzyme requiring light as a physical cofactor to perform its catalytic function. *Biochemistry* 1970, 3, 2610-2620.

Schweikert, K; Hurd, CL; Sutherland, JE; Burritt, DJ. Regulation of polyamine metabolism in *Pyropia Cinnamomea* (W.A. Nelson), an important mechanism for reducing uv-b-induced oxidative damage. *J. Phycol.* 2014, 50, 267-279.

Schweikert, K; Sutherland, J; Hurd, C; Burritt, D. UV-B radiation induces changes in polyamine metabolism in the red seaweed *Porphyra cinnamomea. Plant Growth Regul.* 2011, 65, 389-399.

Sedoud, A; López-Igual, R; Rehman, AU; Wilson, A; Perreau, F; Boulay, C; Vass, I; Krieger-Liszkay, A; Kirilovsky, D. The cyanobacterial photoactive orange carotenoid protein is an excellent singlet oxygen quencher. *The Plant Cell* 2014 tpc.114.123802.

Seeberg, E; Eide, L; Bjørås, M. The base excision repair pathway. *Trends Biochem. Sci.* 1995, 20, 391-397.

Shah, V; Ray, A; Garg, N; Madamwar, D. Characterization of the extracellular polysaccharide produced by a marine cyanobacterium, *Cyanothece* sp. ATCC 51142, and its exploitation toward metal removal from solutions. *Curr. Microbiol.* 2000, 40, 274-278.

Shirkly, B; Kovarcik, DP; Wright, DJ; Wilmoth, G; Prickett, TF; Helm, RF; Gregory, EM; Potts, M. Active Fe-containing superoxide dismutase and abundant sodF mRNA in *Nostoc commune* (cyanobacteria) after years of desiccation. *J. Bacteriol.* 2000, 182, 189-197.

Singh, SC; Sinha, RP; Häder, D-P. Role of lipids and fatty acids in stress tolerance in cyanobacteria. *Acta Protozool.* 2002, 41, 297-308.

Singh, SP; Häder, D-P; Sinha, RP. Cyanobacteria and ultraviolet radiation (UVR) stress: Mitigation strategies. *Ageing Res. Rev.* 2010, 9, 79-90.

Singh, SP; Rastogi, RP; Häder, D-P; Sinha, RP. Temporal dynamics of ROS biogenesis under simulated solar radiation in the cyanobacterium *Anabaena variabilis* PCC 7937. *Protoplasma* 2014, DOI 10.1007/s00709-014-0630-3.

Singh, SP; Rastogi, RP; Sinha, RP; Häder, D-P. Photosynthetic performance of *Anabaena variabilis* PCC 7937 under simulated solar radiation. *Photosynthetica* 2013a, 51, 259-266.

Singh, G; Babele, PK; Sinha, RP; Tyagi, MB; Kumar, A. Enzymatic and non-enzymatic defense mechanisms against ultraviolet-B radiation in two *Anabaena* species. *Process Biochem.* 2013b, 48, 796-802.

Sinha, RP; Häder, D-P. Effects of ultraviolet-B radiation in three rice field cyanobacteria. *J. Plant Physiol.*, 1998,153, 763-769.

Sinha, RP; Lebert, M; Kumar, A; Kumar, HD; Häder, D-P. Disintegration of phycobilisomes in a rice field cyanobacterium *Nostoc* sp. following UV-B irradiation. *Biochem. Mol. Biol. Intern.* 1995a, 37, 697-706.

Sinha, RP; Kumar, HD; Kumar, A; Häder, D-P. Effects of UV-B irradiation on growth, survival, pigmentation and nitrogen metabolism enzymes in cyanobacteria. *Acta Protozool.* 1995b, 34, 187-192.

Sinha, RP; Singh, N; Kumar, A; Kumar, HD; Hader, M; Häder, D-P. Effects of UV irradiation on certain physiological and biochemical processes in cyanobacteria. *J. Photochem. Photobiol. B* 1996, 32, 107-113.

Sinha, RP; Dautz, M; Hader, DP. A simple and efficient method for the quantitative analysis of thymine dimers in cyanobacteria, phytoplankton and macroalgae. *Acta Protozool.* 2001, 40, 187-195.

Sinha, RP; Rastogi, RP; Ambasht, NK; Häder, D-P. Life of wetland cyanobacteria under enhancing solar UV-B radiation. *Proc. Natl. Acad. Sci. India B* 2008, 78, 53-65.

Sinha, RP; Häder, D-P. Biochemistry of phycobilisome disassembly by ultraviolet-B radiation in cyanobacteria. *Recent Res. Dev. Biochem.* 2003, 4945-4955.

Six, C; Joubin, L; Partensky, F; Holtzendorff, J; Garczarek, L. UV-induced phycobilisome dismantling in the marine picocyanobacterium *Synechococcus* sp. WH8102. *Photosynthesis Res.* 2007, 92, 75-86.

Solheim, B; Zielke, M; Bjerke, JW; Rozema, J. Effects of enhanced UV-B radiation on nitrogen fixation in arctic ecosystems. *Plant Ecol.* 2006, 182, 109-118.

Sorrels, CM; Proteau, PJ; Gerwick, WH. Organization, evolution, and expression analysis of the biosynthetic gene cluster for scytonemin, a cyanobacterial UV absorbing pigment. *Appl. Environ. Microbiol.* 2009, 75, 4861-4869.

Soule, T; Stout, V; Swingley, WD; Meeks, JC; Garcia-Pichel, F. Molecular genetics and genomic analysis of scytonemin biosynthesis in *Nostoc punctiforme* ATCC 29133. *J. Bacteriol.* 2007, 189, 4465-4472.

Soule, T; Palmer, K; Gao, Q; Potrafka, RM; Stout, V; Garcia-Pichel, F. A comparative genomics approach to understanding the biosynthesis of the sunscreen scytonemin in cyanobacteria. *BMC Genomics* 2009, 10, 336-345.

Squier, AH; Airs, RL; Hodgson, DA; Keely, BJ. Atmospheric pressure chemical ionisation liquid chromatography/mass spectrometry of the ultraviolet screening pigment scytonemin: characteristic fragmentations. *Rapid Commun. Mass Spectrom.* 2004, 18, 2934-2938.

Srivastava, BS; Kumar, HD; Singh, HN. The effect of caffeine and light on killing of the blue-green alga *Anabaena doliolum* by ultraviolet radiation. *Arch. Microbiol.* 1971, 78, 139-144.

Stock, CA; Dunne, JP; John, JG. Global-scale carbon and energy flows through the marine planktonic food web: An analysis with a coupled physical-biological model. *Prog. Oceanogr.* 2014, 120, 1-28.

Sutherland, DL. Microbial mat communities in response to recent changes in the physiochemical environment of the meltwater ponds on the McMurdo Ice Shelf, Antarctica. *Polar Biol.* 2009, 32, 1023-1032.

Tang, T; Asato, Y. Ultraviolet light induction and photoreactivation of thymine dimers in the cyanobacterium *Anacystis nidulans*. *Arch. Microbiol.* 1978, 118, 193-197.

Taylor, J-S; Lu, HF; Kotyk, JJ. Quantitative conversion of the (6–4) photoproduct of TpdC to its Dewar valence isomer upon exposure to simulated sunlight. *Photochem. Photobiol.* 1990, 51, 161-167.

Tel-Or, E; Huflejt, ME; Packer, L. Hydroperoxide metabolism in cyanobacteria. *Arch. Biochem. Biophys.* 1986, 246, 396-402.

Tichy, M; Vermaas, W. In Vivo role of catalase-peroxidase in *Synechocystis* sp. strain PCC 6803. *J. Bacteriol.* 1999, 181, 1875-1882.

Tomaselli, L; Giovannetti, L. Survival of diazotrophic cyanobacteria in soil. *World J. Microbiol. Biotechnol.* 1993, 9, 113-116.

Tyagi, R; Srinivas, G; Vyas, D; Kumar, A; Kumar, HD. Differential effect of ultraviolet-B radiation on certain metabolic processes in a chromatically adapting *Nostoc*. *Photochem. Photobiol.* 1992, 55, 401-407.

Ueda, T; Kato, A; Kuramitsu, S; Terasawa, H; Shimada, I. Identification and characterization of a second chromophore of DNA photolyase from *Thermus thermophilus* HB27. *J. Biol. Chem.* 2005, 280, 36237-36243.

Van Baalen, C. The effects of ultraviolet irradiation on a coccoid blue-green alga: survival, photosynthesis, and photoreactivation. *Plant Physiol.* 1968, 43, 1689-1695.

Vass, I. Molecular mechanisms of photodamage in the Photosystem II complex. *Biochim. Biophys. Acta* 2012, 1817, 209-217.

Vincent, WF; Quesada, A. Ultraviolet radiation effects on cyanobacteria: implications for Antarctic microbial ecosystems. In: Weiler, CS; Penhale, PA (eds.). Ultraviolet radiation in Antarctica: Measurements and biological effects. Antarctic Research Series, American Geophysical Union, Washington, 1994, pp. 111-124.

Vincent, WF; Neale, PJ. Mechanisms of UV damage to aquatic organisms. In: de Mora, SJ; Demers, S; Vernet, M. (eds.). The Effects of UV radiation on marine ecosystems. Cambridge Univ. Press, Cambridge, 2000, pp. 149-176.

Wachi, Y; Burgess, JG; Iwamoto, K; Yamada, N; Nakamura, N; Matsunaga, T. Effect of ultraviolet-A (UV-A) light on growth, photosynthetic activity and production of biopterin glucoside by the marine UV-A resistant cyanobacterium *Oscillatoria* sp. *Biochim. Biophys. Acta* 1995, 1244, 165-168.

Wang, G; Chen, K; Chen, L; Hu, C; Zhang, D; Liu, Y. The involvement of the antioxidant system in protection of desert cyanobacterium *Nostoc* sp. against UV-B radiation and the effects of exogenous antioxidants. *Ecotoxicol. Environ. Saf.* 2008, 69, 150-157.

Wang, G; Deng, S; Li, C; Liu, Y; Chen, L; Chaozhen Hu, C. Damage to DNA caused by UV-B radiation in the desert cyanobacterium *Scytonema javanicum* and the effects of exogenous chemicals on the process. *Chemosphere* 2012, 88, 413-417.

Werbin, H; Rupert, CS. Presence of photoreactivating enzyme in blue-green algal cells. *Photochem. Photobiol.* 1968, 7, 225-230.

Whitton, BA. Ecology of cyanobacteria II. Their diversity in space and time. Springer, Dordrecht Heidelberg, New York, London, 2012.

Whitehead, K; Hedges, JI. Photodegradation and photosensitization of mycosporine-like amino acids. *J. Photochem. Photobiol. B* 2005, 80, 115-121.

Williams, E; Lambert, J; O'Brien, P; Houghton, JA. Evidence for dark repair of ultraviolet light damage on the bluegreen alga *Gloeocapsa alpicola*. *Photochem. Photobiol.* 1979, 29, 543-547.

Wolfe-Simon, F; Grzebyk, D; Schofield, O; Falkowski, PG. The role and evolution of superoxide dismutases in algae. *J. Phycol.* 2005, 41, 453-465.

Wu, H; Gao, K; Villafañe, VE; Watanabe, T; Helbling, EW. Effects of solar UV radiation on morphology and photosynthesis of filamentous cyanobacterium *Arthrospira platensis*. *Appl. Environ. Microbiol.* 2005, 71, 5004-5013.

Yoshihara, S; Ikeuchi, M. Phototactic motility in the unicellular cyanobacterium *Synechocystis* sp. PCC 6803. *Photochem. Photobiol. Sci.* 2004, 3, 512-518.

Yu, H; Jia, S; Dai, Y. Accumulation of exopolysaccharides in liquid suspension culture of *Nostoc flagelliforme* cells. *Appl. Biochem. Biotechnol.* 2010, 160, 552-560.

Zeeshan, M; Prasad, SM. Differential response of growth, photosynthesis, antioxidant enzymes and lipid peroxidation to UV-B radiation in three cyanobacteria. *South Afr. J. Bot.* 2009, 75, 466-474.

Zhao, W; Guo, Q; Zhao, J. A membrane-associated Mn superoxide dismutase protects the photosynthetic apparatus and nitrogenase from oxidative damage in the cyanobacterium *Anabaena* sp. PCC 7120. *Plant Cell Physiol*. 2007, 48, 563-572.

Reviewed by- Peter Lindblad

In: Cyanobacteria
Editor: Douglas Davison

ISBN: 978-1-61470-925-1

Chapter 7

A SIMPLE CYANOBACTERIA BIOREACTOR DESIGNED FOR UPGRADING PIG FARM BIOGAS

***Yen Lee*[*1], *Jenn-Hung Hsu*[†2], *Yu-Ying Chen*[††3] and *Yueh-Yang Tu*[4#]**

[1]Professor, Department of Life Science,
[2]Associate Professor, Center for General Education,
[3,4]Graduate Students, Department of Life Science,
National Taitung University, Taiwan

ABSTRACT

An easy-to-operate pilot bioreactor designed for biogas H_2S and CO_2 adsorption has successfully upgraded biogas produced from the excreta of 1500 pigs on a farm in rural Taiwan. The bioreactor also seals against unpleasant odors. It costs less than US$200 and can handle at least 3 hours of continuous biogas upgrading, thus contributing to the household gas energy capacity for pig farmers. The bioreactor is designed to dispense a volume of controlled biogas through a small aperture micropore dispenser. The small biogas bubbles pass through water containing cyanobacteria, dissolving the biogas's H_2S and CO_2 into the water, which are then absorbed by cyanobacteria. Key aspects of the design are the amount of water and types of cyanobacteria in the tank, as well as the path length of the bubbles rising through the water. After treatment, the concentration of H_2S in the biogas dropped from 3040 ppm to 2 ppm at the first 10 minutes, 6 ppm after one hour, and arose to 13 ppm after 3 hours. It was acceptably within the upper limit of the threshold for the harmful effects of natural gas (50 ppm). The bioreactor's efficiency could be improved with a longer container for containing the cyanobacteria and a smaller micropore size. The quality of the original biogas could also be upgraded. This system could also lower the CO_2 concentration from 27% to 14% after running for 3 hours.

* Corresponding author: Yen Lee. Tel: +886-913-289-321; Fax: +886-89-356-589. Address: No. 369, Sec.2, University Ave., Taitung 95092, Taiwan. E-mail: yenlee@nttu.edu.tw.
† E-mail: barry@nttu.edu.tw.
†† E-mail: s302972000@yahoo.com.tw.
# E-mail: donicdu@gmail.com.

Keywords: Biogas, H_2S removal, CO_2 removal, Bioadsorption, Cyanobacteria

ABBREVIATION

EPA	Environmental Protection Administration
PE	polyethylene
PVC	Polyvinylchloride

INTRODUCTION

There were 6,126 million pigs in Taiwan according to the last census of agricultural animals and livestock conducted in May of 2010. Taiwan's government encourages pig farm owners to collect and recycle pig excreta into biogas (EPA 2010). Su et al. (2003) analyzed greenhouse gas samples from anaerobic wastewater treatment systems on selected pig farms in Taiwan and estimated that approximately 4,700 tons of CH_4 will be produced every year from Taiwan's pig farms based on census statistics. Currently, the Taiwanese government provides financial subsidies to pig farms for constructing biogas facilities, yet few pig farms have built biogas generators.

The H_2S contained in the biogas is a major barrier to widespread adoption of biogas generators on privately owned and operated farms. Hydrogen sulfide is a highly toxic gas with a rotten egg odor, and it produces SO_2 when burned. The SO_2 forms H_2SO_3 and H_2SO_4 upon reaction with H_2O in air, which then causes iron materials to corrode (Razbani et al. 2011). The majority of pig farms in Taiwan roof their pig houses with corrugated iron sheets. Exposure to sulfuric acid rusts pig house roofs.

Simple economical equipment is needed to reduce H_2S content to acceptable levels in fuel gas. In methane appropriate for pipelines, CO_2 must be below 2% and H_2S must be below 1 ppm (Belmabkhout et al. 2009). Hydrogen sulfide in natural gas may generate corrosive byproducts at concentrations as low as 50 ppm (Bettis 2012). Typically, Taiwanese pig farm owners are unconcerned about the CO_2 concentration of biogas, but they do care about the H_2S concentration.

A study by de Hullu et al. (2008) compared the costs of five methods for removing CO_2 and H_2S from biogas: using iron chelated aqueous amine solutions to adsorb H_2S and CO_2; using a high-pressure water-scrubbing method (in a scrubber, CO_2 and H_2S readily dissolve into water, whereas CH_4 does not); using pressure swing adsorption according to molecular characteristics and the affinity of an adsorption material to separate gases from a mixture of biogas under pressure; using cryogenic separation (different chemicals in biogas liquefy at different temperature and pressure domains); and separating CO_2 and H_2S from CH_4 using a membrane (because of selective permeation). They concluded that the high-pressure water-scrubbing method had the lowest cost at €0.15/m^3, and that this method was both the economical choice and the easiest to operate.

Nevertheless, for gases to be dissolved into liquids, during the gas phase the molecule should be transported to the liquid's surface. Transportation usually occurs by diffusion during the gas phase (Boniface et al. 2000). It is well known that uptake of gaseous H_2S and

CO_2 by aqueous solutions is affected by the temperature and pH (Boniface et al. 2000). Boniface et al. (2000) listed the reaction rates of H_2S, and CO_2 with H_2O:

$$CO_{2(aq)} + 2H_2O <==> HCO_3^- + H_3O^+ \text{at } 291\ °K\ K_1 = (4.0 \pm 0.7) \times 10^3\ M^{-1}\ S^{-1} \quad (1)$$

$$H_2S_{(aq)} + H_2O <==> HS^- + H_3O^+ \text{at } 291\ °K\ K_1 = (1.7 \pm 0.2) \times 10^9\ M^{-1}\ S^{-1} \quad (2)$$

Because H_2S has a saturation concentration in water of approximately 0.125 moles/L (4.25 g/L) at 25°C (Adib et al. 2000) and the CO_2 saturation concentration in water is approximately 0.039 moles/L (1.8 g/L) at 25°C (Garcia 2012), continuous water flow is needed to treat the waste stream. Reducing the cost and improving the convenience of setup and operation are crucial factors for practical usage on pig farms.

Meanwhile, a number of studies have shown how biological components can enhance adsorption in aqueous solutions based on adaptation of photosynthesis in environments containing H_2S. These cyanobacteria can replace oxygenic photosynthesis sensitive to sulfide with an oxygenic photosynthesis that is dependent on sulfide. These cyanobacteria can also use H_2S as an alternative electron donor. Certain cyanobacteria are insensitive to sulfide and oxygenic photosynthetic concurrent with being sulfide-dependent and anoxygenic photosynthetic (Cohen et al. 1986). Some cyanobacteria, in the presence of oxygen, can oxidize H_2S into elemental sulfur, which accumulates in sulfur droplets inside the cell wall. The cyanobacteria can further oxidize this elemental sulfur (Cho et al.1992; Espie et al.1989; Miller and Bebout 2004; Nelson and Castenholz 1981; Strohl and Larkin 1978). Cyanobacteria can also use the dissolved CO_2. Cyanobacteria have the ability to use both CO_2 and HCO_3^- as inorganic carbon sources, as well as ammonia (Markou and Georgakakis 2011). Theoretically, a solution tank containing cyanobacteria could be used to upgrade biogas. Cyanobacteria bioadsorption can degrade H_2S and consume CO_2 that is adsorbed in the water and raises the pH (Ungsethaphand et al. 2009). The cyanobacteria bioadsorption system does not need continuous water flow and should be much easier to operate.

Cohen et al. (1986) estimated that the cyanobacterium *Microcoleus chthonoplastes* at 150 μM H_2S could have a CO_2 photo assimilation rate of 4500 μM min^{-1}μg of chlorophyll a^{-1}. Correspondingly, 1.3 μg of chlorophyll a of *M. chthonoplastes* could be "oxidized" at 585 μM of H_2S per minute. Pandey et al. (2010) grew *Spirulina maxima* in 6 different media, and calculated their dry weights and chlorophyll a contents. The chlorophyll a was approximately 1% of *S. maxima*'s dry weight. Estimating from these data, 0.13 mg of *M. chthonoplastes* might consume 585μM of H_2S per minute and photosynthesize 5850 μM of CO_2. Espie et al. (1989) treated *Synechococcus* UTEX 625 cells with up to 150μM H_2S + HS^- at pH 8.0. This treatment had little effect on HCO_3^- transport dependent on Na^+ or photosynthetic O_2 evolution, but CO_2 transport was inhibited by more than 90%. For cyanobacteria, intercellular carbon is in the form of HCO_3^-. HCO_3^- is converted to CO_2 by the enzyme carbonicanhydrase according to the reaction: $CO_2 + H_2O \leftrightarrow HCO_3^- + H^+$ (Markou and Georgakakis 2011). The photosynthesis is not inhibited. The physical reason is that biodegradation reactions take time and require enough water contention, cells, and surface contact. If biogas releases into a water tank containing cyanobacteria while tiny bubbles slowly emerge from a long porous tube, H_2S and CO_2 have sufficient time and concentration for liquid transition phase surface contact to dissolve into the water. The cyanobacteria may eliminate H_2S and CO_2 in the aqueous phase.

The length of the porous tube, pore size, amount of water, and amount of cyanobacteria would be the parameters affecting the efficiency.

In this report, a simple set up used cyanobacteria bioadsorption to remove H_2S and CO_2 from biogas. This invention could improve the quality of biogas and promote the use of bio-renewable energy on pig farms.

MATERIALS AND METHODS

Biogas Resource

This experiment used biogas from FongYi Pig Farm in Taitung County, Taiwan. FongYi Pig Farm has approximately 1500 pigs. The farm used the pigs' dung, urine, and pig house wastewater to produce biogas. Biogas was collected in a polyethylene (PE) plastic bag, and a 4.5cm diameter_Polyvinylchloride (PVC) pipe delivered the biogas to the owner's house for the family's daily gas energy supply. A 1/8 horsepower air pump was hooked to the outlet of the pipe and volume valve controlled delivery system (1.5 L/min). The delivery system was connected to the plastic tube attached to the micropore air dispenser.

Hydrogen Sulfide Gas Adsorption Bioreactor System

A transparent plastic tube 50 cm long and 0.8 cm in diameter was connected to the outlet attached to a micropore air dispenser (Brand: Mr. Aqua, 15 cm in length, bobble size 100 to 500 μm) and inserted into each of the two 35 L transparent PVC water tanks containing cyanobacteria (50 cm in height, 30 cm in diameter). The water tanks absorbed the H_2S and CO_2, as shown in Figure 1 and the schematic diagram shown in Figure 2.

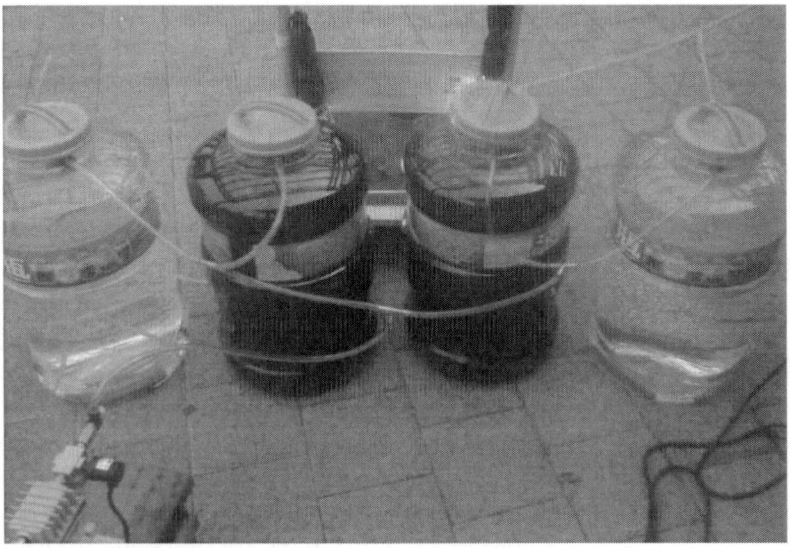

Figure 1. A model of the hydrogen sulfide bioreactor adsorption system.

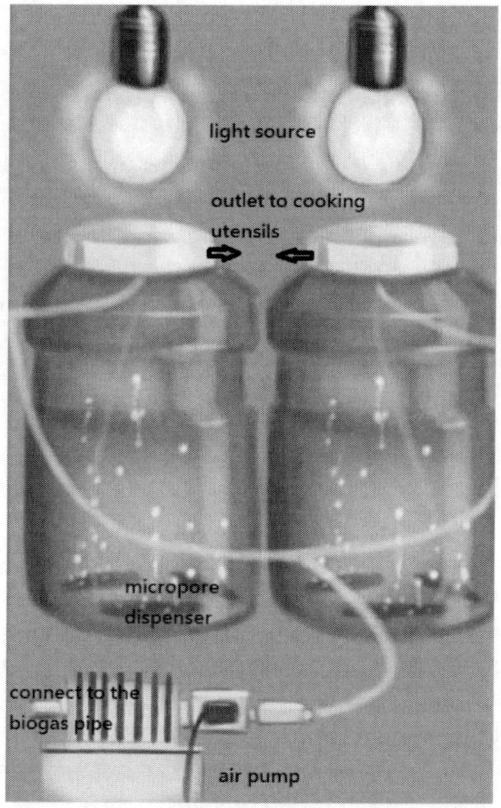

Figure 2. The schematic diagram of the hydrogen sulfide bioreactor adsorption system.

The cyanobacteria treatment tanks were filled with tap water and 87.5 g (wet weight) *Lyngbya* sp. mixed with 87.5 g (wet weight) *Cyanothece* sp. The cyanobacteria were dominant strains. Lee and Lee (2009) separately isolated these two cyanobacteria strains from the Beinan River and the Taiping River (both located in Taitung County, Taiwan). A 50 cm transparent plastic tube touched the bottom of each tank. The outlet end of the tube was connected to a micropore air dispenser. The biogas was pumped (1.5 L/min) from the pig farm biogas pipe into the inlet of the tube, and the biogas bubbled in the tanks. The H_2S and CO_2 in the biogas dissolved in water and were used by the cyanobacteria. The unabsorbed biogas was released from the other outlet on the tank top and delivered to the kitchen stove. The control set used the same type of design as described, but the tanks were filled with tap water that did not have cyanobacteria.

Cyanobacteria Culture

Equal weights of *Lyngbya* sp. and *Cyanothece* sp. were mixed. The dominant strain was used. These two cyanobacteria strains were previously separately tested for H_2S tolerance by delivering biogas into their growth tanks continuously for 8-hrs/day for 7 days. Thereafter, their cells were microscopically checked (data not shown). No harmful effects on the cells were found. Each strain was separately maintained in 20 L transparent covered plastic tanks in a BG-11 medium (Allen and Stanier, 1968), as described by Li and Lee (2012). Cells (wet weight) were collected using the suction of a 1/8 horsepower pump to filter the culture media through a piece of filter paper on a funnel. Equal amounts (87.5 g each) of wet weight *Lyngbya* and *Cyanothece* cells were homogenized by a polytron before adding the cyanobacteria to the water tank. The cyanobacteria were suspended by the lifting effect of the bubbling gas. The system was set in an area exposed to sunlight by day and lit with a >7400 lux light bulb during the night for continuous illumination. When the biogas is not in use, the 1/8 horsepower pump should be disconnected and the system switched to a small aquarium air pump (this study used a pump of 110V, 4.5W, and 50 Hz for approximately 4400 mL of air pumped per minute) to pump air into the bioreactor, to lift the cyanobacteria and provide air for cells to photosynthesize and ready the bioreactor for future use.

Biogas Amount, and H_2S, NH_3, CO_2, R-NH_2, As Well As CH_4 Concentrations Detection

The water displacement method was used to measure the volume of biogas. A Gastec GV100 (GASTEC Corporation, Japan) hand-sampling pump equipped with a detector tube was used for sampling gases as described by Li and Lee (2012). The gas samples were collected at 10 minute, 60 minute, and 180 minute intervals after the biogas was delivered into the tank. When collecting the air samples, the tip of the detector tube was inserted in the outlet tube, and gas was pumped according to the manufacturer's instructions. The reading was adjusted for the temperature and related compensation factors as described in the Gastec operation manual. Detector tubes were chosen to fit their sensitivity ranges. After sucking air into the detector tube, the gas in the air reacts with the chemical inside the tube. Reading the length of the color change of the gas inside the tube reveals the concentration of a specific gas. Gasteck type 4LT (detects H_2S, measurement range 0.1 to 4 ppm), type 4HM (detects H_2S, sensitivity 25 to 1600 ppm), type 4L (detects H_2S, sensitivity 1 to 240 ppm), type 3La (detects NH_3, sensitivity 2.5 to 200 ppm), type 180L (detects R-NH_2, sensitivity 0.5 to10 ppm), type 2L (detects CO_2, sensitivity 0.13% to 6.0%), and type 2HH (detects CO_2, sensitivity 2.5% to 40%) measuring tubes separately detected H_2S, NH_3, R-NH_2, and CO_2 in the biogas. A GasAlerMax XT (1 to 100%) handheld instrument was used to measure CH_4 concentration. Triplicate measurements were made for each test and the averages were presented. The H_2S concentration of the original biogas was higher than the upper limit of the 1600 ppm detection range. To compensate for this limitation, an adjusted sample size of 25 mL was sampled each time and the measured number was multiplied by four.

pH Values, Water Temperature

The pH meter (Milwaukee pH 600) measured the liquid pH values before and after biogas delivery. The pH values of the water tanks containing cyanobacteria (the bioreactor) were also measured 12 hours after the initial treatment to check the cyanobacteria's recovery from H_2S and CO_2 adsorption.

Thermometers were inserted in each tank about 10 cm deep and measured the water temperature in both the experimental and control tanks before and after the experiment.

Statistical Analysis

Data were reported by calculating the means from duplicated experiments. Paired *t*-tests were used to compare control group H_2S and CO_2 concentration averages with the experimental group values, respectively. A confidence level of 95% was chosen.

RESULTS

Biogas Delivery

A 1/8 horsepower pump delivered the biogas from the pig farm biogas generator pipe, and a control valve evenly distributed the biogas into the experimental tanks with cyanobacteria or the control water tanks. The amount of biogas delivered into each tank was counted at 1.5 L/min.

H_2S, NH_3, R-NH_2, CO_2, and CH_4 concentrations

The average measurements of the H_2S, NH_3, R-NH_2, CO_2, and CH_4 concentrations of the original biogas from the storage tank, the first 10 minutes after the treatment, 60 minutes and 180 minutes after the treatment, separately for both the water control tanks and cyanobacteria containing experimental groups, are listed in Table 1.

Table 1. The H_2S, NH_3, R-NH_2, CO_2, and CH_4 averaged concentrations measured before and after the treatments

Location of Concentration/measurement		H_2S (ppm)	NH_3 (ppm)	CO_2 (%)	R-NH_2 (ppm)	CH_4 (%)
Original biogas from the storage tank		3040	Not detected	27	Not detected	64
Water control bottle outlet	10 min	12	Not detected	10	Not detected	83
	60 min	85	Not detected	17	Not detected	73
	180 min	280	Not detected	17	Not detected	74
Cyanobacteria containing bottle outlet	10 min	0	Not detected	4	Not detected	88
	60 min	6	Not detected	8	Not detected	83
	180 min	13	Not detected	14	Not detected	78

The NH_3 and R-NH_2 concentrations were not detected in the research period by the Gasteck type 3La (sensitivity 2.5-200 ppm) and type 180L (sensitivity 0.5-10 ppm) separately, whether before or after the treatment from the water tanks or the cyanobacteria containing tanks.

Data Analysis

The H_2S concentration decreased rapidly in both the water control tanks and cyanobacteria containing tanks (Figure 3). The *p* value of the paired *t*- test is 0.12, indicating no significant difference between the control and experimental treatment.

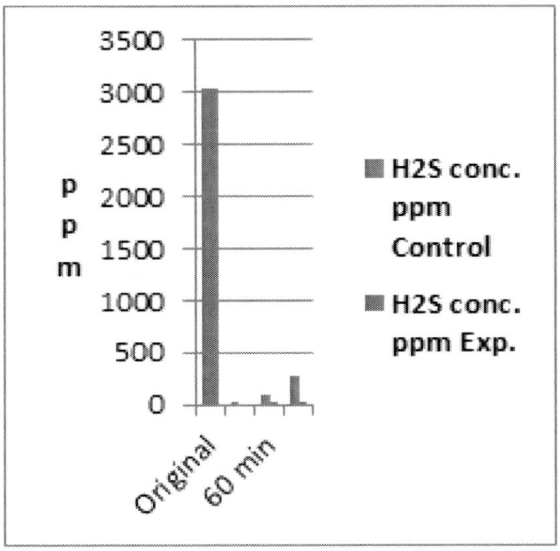

Figure 3. The average H_2S concentrations measured from control and cyanobacteria containing tanks during the experimental period.

The CO_2 concentration decreased in both the water control tanks and cyanobacteria containing tanks (Figure 4). The *p* value of the paired *t*-test is 0.05, indicating that the experimental tanks are more efficient.

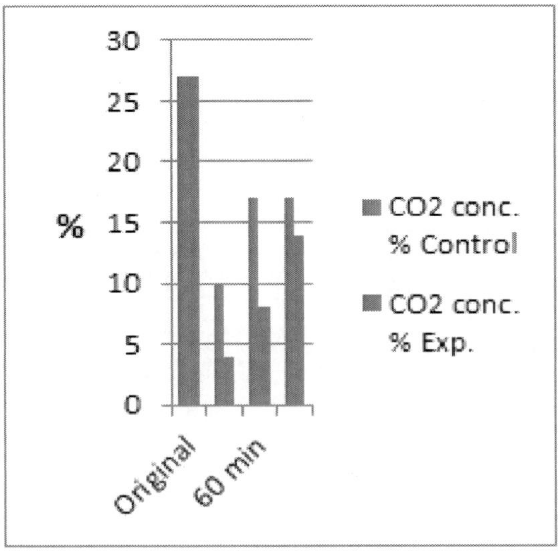

Figure 4. The average CO_2 concentrations measured from control and cyanobacteria containing tanks during the experimental period.

The CH_4 concentration increased at the first 10-minute interval and then gradually reduced in both the water control tanks and cyanobacteria containing tanks (Fig. 4). The *p* value of the paired *t*- test is 0.05, indicating that the experimental tanks are more efficient.

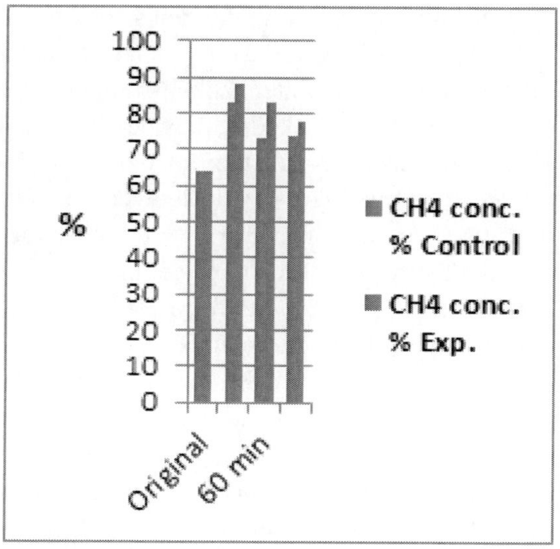

Figure 5. The average CH_4 concentrations measured from control and cyanobacteria containing tanks during the experimental period.

pH values

The liquid pH value of the control tanks before treatment was 8.0. After three hours of treatment, the liquid pH value of the control tanks changed to 6.1. Before treatment, the pH value of the water tank containing cyanobacteria was 8.8, and after three hours of treatment the pH value changed to 7.2. Twelve hours later, after air bubble lifting and regular photosynthesis, the pH value of the bioreactor was 8.4.

Water temperature

The liquid temperature of the control tanks before treatment was 29°C, and after three hours of treatment the temperature showed no change. The temperature of the water tanks containing cyanobacteria before treatment was 29°C, and after three hours of treatment the temperature showed no change.

CONCLUSION

This is the first study of upgrading a biogas system (to reduce H_2S and partially reduce CO_2 concentration) designed for Taiwan's small-to-midsize pig farms (<1500 pigs). This study is significant primarily because it opens a route for swine farmers in Taiwan or even around the world as an easy method of upgrading biogas produced from pig excreta, without worrying about the corrosion of their iron pigpen roofing or metal tools. Also, this research can encourage them to build biogas fermenters, and save the valuable bio-renewable energy, as well as reduce the global warming effect. The upgrade design is economical, easy to operate, and the system is sealed against bad odors. The system was both successfully constructed and operated.

Secondly, the research results showed that the water tanks and the cyanobacteria containing tanks both had high efficiency to absorb H_2S and CO_2, and upgrade CH_4 concentration, although the cyanobacteria containing tanks showed greater efficiency in CO_2 adsorption, and upgrading CH_4 than the water tanks, yet for those farmers unfamiliar with cyanobacteria handling, they still can use water tanks to upgrade the biogas but they need slowly running water inject into the tanks to keep the water fresh and maintain the function. The advantages of the cyanobacteria tanks are they can function longer and re-fresh their function by lighting and air-bubbling. The study showed that at 60 minutes, the H_2S concentrations released from the control water tanks were higher than the upper limit of the threshold to generate corrosive byproducts (50 ppm) (Bettis 2012), while the cyanobacteria containing tanks kept 13 ppm of the H_2S even with release after 180 minutes.

Finally, the cyanobacteria used in this system are easy to grow and maintain in tap water or ground water tanks. Any farmer can learn the culture technique and install the biogas upgrading system in one day without spending large amounts of money and weeks to learn construction of biogas purification systems.

In this study, dominate cyanobacteria strains were used. The reason for mixing two strains was that pre-tests occasionally found that any one of *Lyngbya* or *Cyanothece* used in the system would yield extensive contamination by different cyanobacteria. However, when two of these strains are mixed, the contamination was reduced (data not shown).

That the added total concentration of gases did not add up to 100% might be because of the deviation of the GasAlerMax XT instrument, and a certain percentage of the concentration was comprised of other gases such as H_2 and CO.

The H_2S and CO_2 concentrations showed a gradual rise in both the control and water tanks containing cyanobacteria. This phenomenon suggests that H_2S and CO_2 dissolved in water or water tanks containing cyanobacteria will gradually become saturated. The cyanobacteria could also consume part of the dissolved H_2S and CO_2, causing saturation to be slow. Light is an important factor in cyanobacteria photosynthesis. When the biogas was delivered through the cyanobacteria tank without a light source, the H_2S and CO_2 concentrations were similar to the control tank (data not shown). During nighttime, the system needs a >7400 lux light source to operate. This study used Philips 23W Helix lamps, and together with pump energy consumption, the apparatus could add extra costs to individuals who adopt this system.

Another key point is the amount of biogas delivery. For this size of cyanobacteria bioreactor, the 1.5 L/min amount was adequate for at least 3 hours of running. Three hours of biogas supply should be enough for a family's cooking and shower usage needs, and 1.5 L/min of biogas is enough for a gas stove.

The CO_2 and H_2S must be dissolved in water to be used by the cyanobacteria. The biogas bubbled out through the micropore air dispenser from the bottom of the liquid tank. The bubble retention time and bubble size could affect the quantity of CO_2 and H_2S dissolving in the water. During this research period, the 100 to 500 μm bubble size and 35 L liquid tank (50 cm in length) proved to be adequate. The molecular diameter of CO_2 is approximately 0.34 nm (Huang et al., 2007), and H_2S has a diameter of approximately 4.0 Å (Kammeyer and Whitman, 1972). It is reasonable to have a pore size of 100 to 500 μm because the gas should have a large contact surface area with the water. If needed, a longer length of the tube, smaller pore size, or larger tank (with more water and cyanobacteria) could be used.

The pH values were lowered in both control water tanks and water tanks containing cyanobacteria. The temperature values showed no change before and after treatment in both control tanks and water tanks containing cyanobacteria. When the biogas stopped running and cyanobacteria started regular photosynthesis, the system returned to pH 8.4 in 12 hours, and the cells were ready for another cycle of the bioreactor. An extra bioreactor may be needed for alternating usage.

It is difficult to promote biogas systems on Taiwan pig farms, but the biogas upgrading system is easy to operate and financially affordable. Ease of operation and financial affordability were the primary purposes of this design.

ACKNOWLEDGMENTS

Sincere appreciation is extended to FongYi Pig Farm in Taitung County, Taiwan, for providing facilities and support during the execution of this project. The authors also thank Mr. Ch'ao, Ch'un-Cheng in National Taitung University (NTTU) for drawing the system's schematic diagram (Figure 2) and Mr. Chen, Kuan-Da as well as Mr. Jo, Shan-Huang, both at NTTU, for help in taking measurements.

REFERENCES

Adib F, Bagreev A, Bandosz TJ. Analysis of the relationship between H_2S removal capacity and surface properties of unimpregnated activated carbons. *Environ Sci Technol.* 2000; 34:686–692.

Allen MM, Stanier RY. Growth and division of some unicellular blue-green algae. *J Gen Microbiol.* 1968; 51:199–202.

Belmabkhout Y, Weireld GD, Sayari A. Amine-bearing mesoporous silica for CO_2 and H_2S removal from natural gas and biogas. *Langmuir Letter* 2009; 25:13275–13278.

Bettis. Molecular sieve desiccant dehydrator for natural gas [Internet]. 2012 [cited 2013 Sep 6]. Available from: http://www2.emersonprocess.com/siteadmincenter/PM%20Valve%20Automation%20Documents/Bettis/Brochure/MolecularSieve.pdf.

Boniface J, Shi Q, Li YQ, Chung JL, Rattigan OV, Davidovits P. Uptake of Gas-phase SO_2, H_2S, and CO_2 by aqueous solutions. *J Phys Chem.* 2000; 104:7502–7510.

Cho KS, Hirai M, Shoda M. Degradation of hydrogen sulfide by *Xanthomonas* sp. Strain DY44. isolated from peat. *Appl Environ Microbiol.* 1992; 58: 1183–1189.

Cohen Y, Jorgensen B, Revsbech NP, Poplawskil R. Adaptation to hydrogen sulfide of oxygenic and anoxygenic photosynthesis among cyanobacteria. *Appl Environ Microbiol.* 1986; 51:398-407.

Environmental Protection Administration Report[Internet]. Create Sustainable Rural Villages Through Clean Pig Farming and Renewable Green Energy. 2010 [cited 2013 Jul 3]. Available from: http://www.businesswire.com/news/home/20101223005871/en/EPA-Create-Sustainable-Rural-Villages-Clean-Pig.

Espie GS, Miller AG, Canvin DT. Selective and reversible inhibition of active CO_2 transport by hydrogen sulfide in a cyanobacterium. *Plant Physiol.* 1989; 91: 387–394.

Garcia MA. Garcia's chemical Forums [Internet]. 2012 [cited 2012 Dec 25]. Available from: http://www.chemicalforums.com/index.php?topic=43230.0

Huang LL, Zhang LH, Shao Q, Lu LH, Lu XH, Jiang SY, et al. Simulations of binary mixture adsorption of carbon dioxide and methane in carbon nanotubes: temperature, pressure, and pore size effects. *J Phys Chem*. 2007; C111:11912–11920.

de Hullu J, Maassen JIW, van Meel PA, Shazad S, Vaessen JMP, Bini LJC, et al. Comparing different biogas upgrading techniques. *Final Report*, Eindhoven University of Technology. 2008; 1–36.

Kammeyer CW, Whitman DR. Quantum mechanical calculation of molecular radii. I. hydrides of elements of periodic groups IV through VII. *J Chem Phys*. 1972; 56: 4419–4421.

Lee CL, Lee Y. Cyanobacterial bio-indicator survey for two main rivers in Taitung Taiwan. *J Ecol Environ Sci.* 2009; 2:1–26.

Li CS, Lee Y. Household cyanobacteria bio-reactor to diminish kitchen waste sewage malodor and produces fertilizer. *Int J Appl Sci Eng*. 2012; 10:29–39.

Markou G, Georgakakis D. Cultivation of filamentous cyanobacteria (blue-green algae) in agro-industrial wastes and wastewaters: A review. *Appl Energ*. 2011; 88:3389–3401.

Miller SR, Bebout BM. Variation in sulfide tolerance of photosystem II in phylogenetically diverse cyanobacteria from sulfidic habitats. *Appl Environ Microbiol*. 2004; 70:736–744.

Nelson DC, Castenholz RW. Use of reduced sulfur compounds by *Beggiatoa* sp. J Bacteriol. 1981; 147:140–154.

Pandey JP, Tiwari A, Mishra RM. Evaluation of biomass production of *Spirulina maxima* on different reported media. *J Algal Biomass Utln.* 2010; 1:70–81.

Strohl WR, Larkin JM. Enumeration, isolation, and characterization of *Beggiatoa* from freshwater sediments. *Appl Environ Microbiol*. 1978; 36:755–770.

Su JJ, Liu BY, Chang YC. Emission of greenhouse gas from livestock waste and wastewater treatment in Taiwan. *Agric Ecosyst Environ*. 2003; 95:253–263.

Razbani O, Mirzamohammad N, Assadi M. Literature review and road map for using biogas in internal combustion engines. In: *Third International Conference on Applied Energy,* Perugia, Italy; 2011.

Ungsethaphand T, Peerapornpisal Y, Whangchai N. Production of *Spirulina platensis* using dry chicken manure supplemented with urea and sodium bicarbonate. *Maejo Int J Sci Technol.* 2009; 3:379–387.

Reviewed by: Dr. Shang-Lien Lo
Affiliation: Professor, Graduate Institute of Environmental Engineering,
National Taiwan University.
71 Chou-Shan Rd., Taipei 106, Taiwan. E-mail: sllo@ntu.edu.tw,
Tel: 886-2-2362-5373 , 886-2-3366-4377

In: Cyanobacteria
Editor: Douglas Davison

ISBN: 978-1-61470-925-1

Chapter 8

CELL-PENETRATING PEPTIDE-MEDIATED PROTEIN UPTAKE IN CYANOBACTERIA

Betty Revon Liu[1], Yue-Wern Huang[2], Robert S. Aronstam[2] and Han-Jung Lee[1,*]

[1]Department of Natural Resources and Environmental Studies, National Dong Hwa University, Hualien, Taiwan
[2]Department of Biological Sciences, Missouri University of Science and Technology, Rolla, MO, US

ABSTRACT

Cyanobacteria are photosynthetic bacteria that have played important roles in the development of the Earth and its atmosphere. Accordingly, novel experimental techniques for modifying their regulation and function are of great interest. Cell-penetrating peptides (CPPs) have attracted scientists' attention in recent decades, because they can overcome the hydrophobic plasma membrane barrier and enter cells directly. This phenomenon is known as protein transduction. In this study, we demonstrate that one CPP, the nona-arginine (R9) peptide, is able to deliver noncovalently associated green fluorescent protein (GFP) into *Synechococcus elongatus* PCC 7942 cyanobacteria. Reduction of GFP internalization by physical and pharmacological inhibitors demonstrated that uptake of uncomplexed GFP by cyanobacteria is mediated by classical endocytosis. CPP-mediated protein transduction of R9/GFP complexes was inhibited by treatment with N-ethylmaleimide (NEM), and macropinocytic inhibitors diminished CPP-mediated protein transduction of R9/GFP complexes in either the absence or presence of NEM. These two lines of evidence suggest that the major mechanism for CPP-mediated protein transduction is macropinocytosis, and classical endocytosis plays a minor role. Collectively, our CPP investigations offer the insight of the understanding of protein import in cyanobacteria.

* Corresponding author: Han-Jung Lee. E-mail: hjlee@mail.ndhu.edu.tw.

Abbreviations

CPP	cell-penetrating peptide
CytD	cytochalasin D
EIPA	5-(*N*-ethyl-*N*-isopropyl)-amiloride
GFP	green fluorescent protein
NEM	*N*-ethylmaleimide
PTD	protein transduction domain
R9	nona-arginine
Tat	transactivator of transcription
Wort	wortmannin

Introduction

Cyanobacteria (blue-green algae) were first described as a distinct group of microorganisms by the Breslaw botanist Ferdinand Cohn, who initially classified them as "bacterialike organisms" [1]. Cyanobacteria are photo-autotrophic prokaryotes that perform oxygenic photosynthesis [2]. Cyanobacteria contribute significantly to the maintenance of atmosphere architecture and play an essential role in carbon cycles and oxygen production [2]. Their phototropic metabolisms are thought to have provided the basis for chloroplast function in plants through an endosymbiotic process [3]. Recently, cyanobacteria have emerged as prime candidates for industrial applications, including the production of biofuels and fine chemicals [4].

Cell-penetrating peptides (CPPs), also termed protein transduction domains (PTDs), are short peptides that possess the ability to penetrate plasma membranes [5]. The first CPP, transactivator of transcription (Tat), was discovered in a transcriptional transactivator of the immunodeficiency virus type 1 (HIV-1) by two groups in 1988 [6, 7]. Subsequently, many additional CPPs were discovered with common compositions of cationic, amphipathic, or hydrophobic amino acids [8, 9]. Their ability to penetrate plasma membranes offered potential applications in genetic engineering and pharmaceutical biology.

Not only can CPPs, such as nona-arginine (R9) peptides, enter cells by themselves, but they can also deliver various cargoes into cells [5]. A strikingly wide variety of cargoes can be delivered by CPPs, including bioactive proteins, DNAs, RNAs, cytotoxic therapeutic drugs, and inorganic nanoparticles [8–10]. The advantages of CPPs as carriers of therapeutic agents include low toxicity, rapid transduction rate, ease of use, and high penetration efficiency [11]. However, the mechanisms underlying CPP-mediated cellular internalization remain controversial. Factors that influence the entry routes of CPPs into cells include amino-acid compositions of the CPPs, CPP concentrations, cell type, zeta potential of the CPP-cargo complexes, and particle size [12]. To date, two major routes for CPP penetration have been identified: direct membrane translocation and endocytosis-mediated entry [5, 13]. Endocytic mechanisms can be further resolved into clathrin-dependent endocytosis, caveolae-dependent endocytosis, and nonclassical macropinocytosis [14].

In previous studies, we demonstrated that macropinocytosis is the major route for cellular internalization of R9 or noncovalently associated R9/cargo complexes [15]. In contrast, histidine-modified arginine-rich CPPs (HR9) were found to deliver semiconductive nanomaterials into cells by direct membrane translocation [16]. More recently, we demonstrated that cyanobacteria use classical endocytosis and macropinocytosis to internalize exogenous GFP and R9/GFP proteins, respectively [17]. Accordingly, CPPs which utilize different internalization pathways are valid tools for studies of cellular uptake of materials which affect diverse cellular processes. In this report, we studied transduction efficiency and mechanisms of cellular entry of R9/GFP complexes in *Synechococcus elongatus* PCC 7942 cyanobacteria.

MATERIALS AND METHODS

Culture of Cyanobacteria

Synechococcus elongatus PCC 7942 (American Type Culture Collection, Manassas, VA, USA) was cultured in BG-11 medium with mild agitation at 28°C, as previously described [17].

Plasmid and Protein Preparation

The pR9 plasmid consists of a coding sequence of R9 under the control of the T7 promoter [15]. The pQE8-GFP plasmid contains a coding sequence for green fluorescent protein (GFP) under the control of the T5 promoter [18]. Plasmid DNA was prepared and purified using the Nucleobond AX100 Kit (Machery-Nagel, Duren, Germany). For protein expression, plasmids were individually transformed into *Escherichia coli*, as previously described [15, 18]. Expressed proteins were then purified, concentrated, and quantified using the Protein Assay Kit (Bio-Rad, Hercules, CA, USA).

Noncovalent Protein Transduction

Cyanobacteria were treated with R9 or R9/GFP complexes, as previously described [17]. To prepare the complexes, R9 peptide was incubated with GFP at a molecular ratio of 3:1 at room temperature for 10 min. To investigate the internalization of exogenous proteins, cyanobacteria were washed with double-deionized water and treated with BG-11 medium as a negative control. Cells were exposed to either uncomplexed GFP at a final concentration of 800 nM or noncovalent R9/GFP complexes prepared at a molecular ratio of 3:1.

To identify the mechanism of protein internalization in cyanobacteria, physical and pharmacological inhibitors of membrane transport processes, including low temperature (4°C), 2 μM of valinomycin (Sigma-Aldrich, St. Louis, MO, USA), 2 μM of nigericin (Fluka Chemie, Seelze, Germany), and 2 mM of *N*-ethylmaleimide (NEM, Sigma-Aldrich), and 10 μM of fusicoccin (Sigma-Aldrich), were used, as previous described [17, 19]. To study macropinocytosis, cyanobacteria were treated with 100 μM of 5-(*N*-ethyl-*N*-isopropyl)-amiloride (EIPA, Sigma-Aldrich), 10 μM of cytochalasin D (CytD, Sigma-Aldrich), or 100 nM of wortmannin (Wort, Sigma-Aldrich), followed by the exposure to R9/GFP complexes [17, 19].

Fluorescent and Confocal Microscopy

Fluorescent and bright-field images were obtained using an Olympus BX51 fluorescent microscope (Olympus, Center Valley, PA, US) or an Eclipse E600 fluorescent microscope (Nikon, Melville, NY, USA), as previously described [17]. Confocal images were obtained using a TCS SL confocal microscope system (Leica, Wetzlar, Germany), as previously described [17]. GFP was detected using excitation at 488 nm and emission at 500-530 nm. Transduction efficiency was determined from the digital image data as analyzed using the UN-SCAN-IT software (Silk Scientific, Orem, UT, US) [19].

Statistical Analysis

Results are expressed as mean ± standard deviation (SD). Mean values and SDs were calculated from at least three independent experiments carried out in triplicates for each treatment group. Statistical comparisons were performed by ANOVA, using levels of statistical significance of $P < 0.05$ (*, α, and β) and $P < 0.01$ (**, αα, and ββ), as indicated.

RESULTS

To assess protein internalization in cyanobacteria, cells were treated with either uncomplexed GFP or R9/GFP complexes. Exogenous GFP and R9/GFP complexes were both delivered into cyanobacteria (Figure 1).

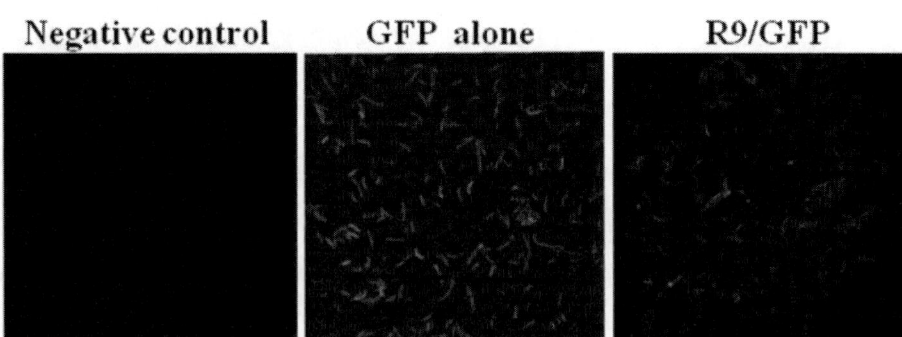

Figure 1. Protein internalization in *Synechococcus elongatus* PCC 7942. Cyanobacteria were treated with either 800 nM of GFP or R9/GFP complexes prepared at a molecular ratio of 3:1. Cells were washed with double-deionized water and treated with BG-11 medium as a negative control. Images of GFP channel were recorded using a TCS SL confocal microscope system (Leica, Wetzlar, Germany) at a magnification of 1,000 ×.

To investigate the mechanism of proteins uptake in cyanobacteria, the cells were treated with GFP in the presence or absence of physical and pharmacological inhibitors of classical endocytosis, including NEM, fusicoccin, nigericin, valinomycin, low temperature (4°C), and sodium azide (Figure 2A). Endocytic efficiency of GFP was significantly reduced in cyanobacteria treated with each of these inhibitors (Figure 2B). Among these inhibitors, NEM was the most effective at blocking endocytosis. Accordingly, treatment with NEM was used in subsequent experiments. These results are consistent with our previous data [17], and indicate that classical endocytosis is the major route for internalization of uncomplexed GFP in cyanobacteria.

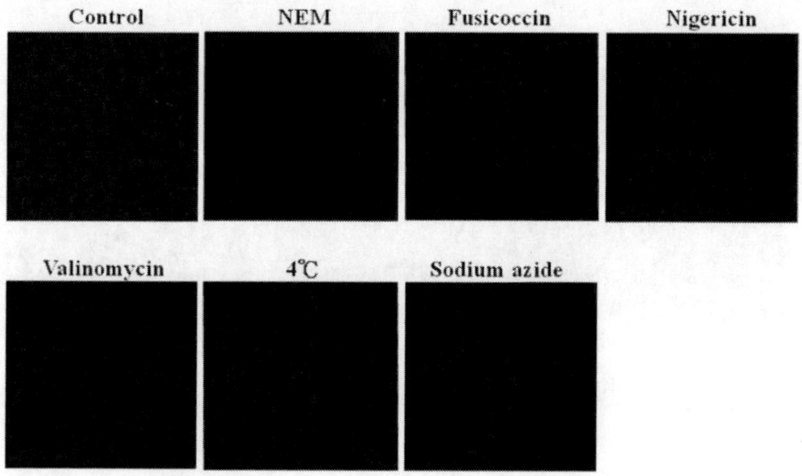

A

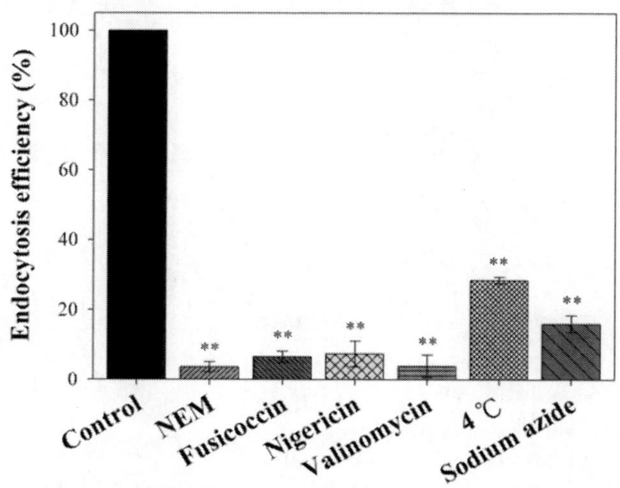

B

Figure 2. The mechanism of GFP internalization in cyanobacteria. (A). Endocytic pathway. Cyanobacteria were treated with 800 nM of GFP either in the absence of endocytic inhibitors (Control) or in the presence of NEM, fusicoccin, nigericin, valinomycin, low temperature (4°C), and sodium azide, as indicated. (B). Histogram of endocytic efficiency. The results depicted in Figure 2A were quantified using the UN-SCAN-IT software. Significant differences (*, $P < 0.05$; **, $P < 0.01$) are indicated. Data are presented as mean ± SD from five independent experiments in each treatment group.

To study whether CPP-mediated protein transduction also involves endocytosis, cyanobacteria were treated with either uncomplexed GFP or R9/GFP complexes in the presence or absence of NEM (Figure 3A). Although NEM treatment decreased endocytic efficiency in cyanobacteria treated with uncomplexed GFP or R9/GFP complexes, approximately 50 and 76% of protein internalization remain, respectively (Figure 3B). In addition, endocytic efficiency of R9/GFP complexes was superior to uncomplexed GFP in cyanobacteria treated with NEM. These results indicated that classical endocytosis is one of the mechanisms for both uncomplexed GFP and CPP-mediated protein transduction. Therefore, we hypothesize that, other than classical endocytosis, there is still an alternative major route of protein transduction in cyanobacteria.

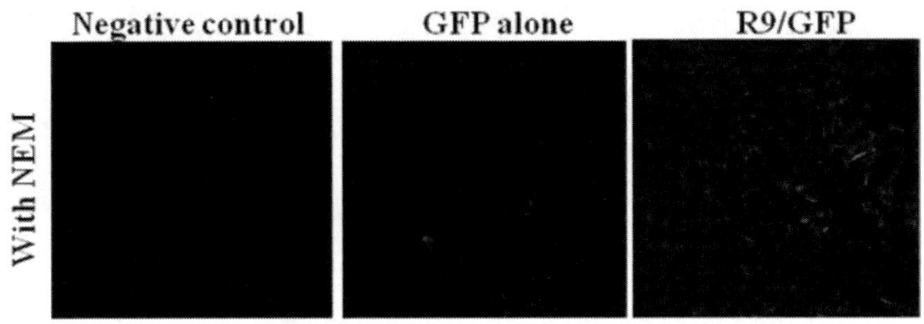

A

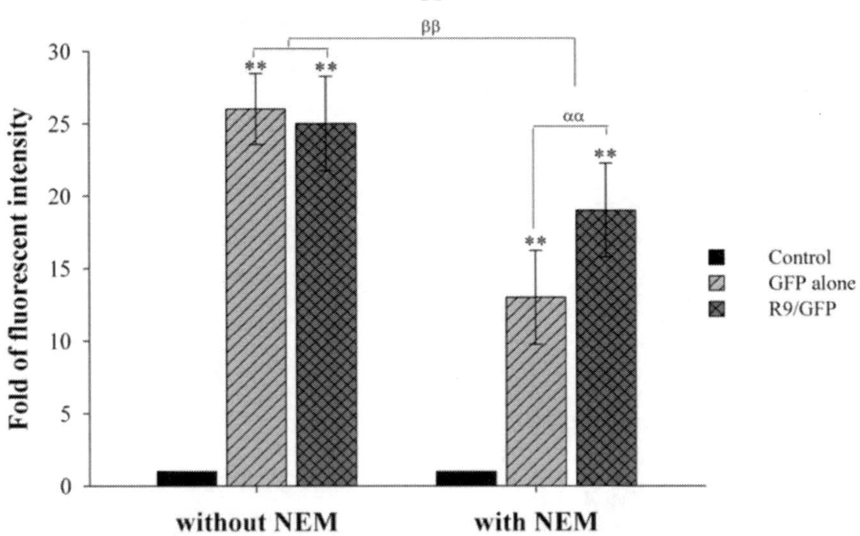

B

Figure 3. The mechanism of CPP-mediated protein transduction in cyanobacteria. (A). Protein transduction in cyanobacteria was measured in the absence or presence of NEM. Cyanobacteria were treated with either 800 nM of GFP or R9/GFP complexes in the absence (Control) or presence of 2 mM of NEM. Cells were observed at a magnification of 1,000 × using a TCS SL confocal microscope. (B). Effects of NEM on protein internalization. Results from Figure 3A were quantified using the UN-SCAN-IT software. Significant differences of $P < 0.05$ (*, α, β) and $P < 0.01$ (**, αα, ββ) are indicated. Data are presented as mean ± SD from five independent experiments in each treatment group.

To identify the major route, other than classical endocytosis, for cellular entry of R9/GFP complexes in cyanobacteria, cells were treated with R9/GFP complexes in the presence or absence of NEM and macropinocytic inhibitors, including CytD, EIPA, and Wort, as indicated (Figure 4A). All of these macropinocytic inhibitors significantly reduced protein transduction in cyanobacteria in the presence and absence of NEM (Figure 4B). These results indicate that protein transduction of R9/GFP complexes in cyanobacteria involves lipid raft-dependent macropinocytosis.

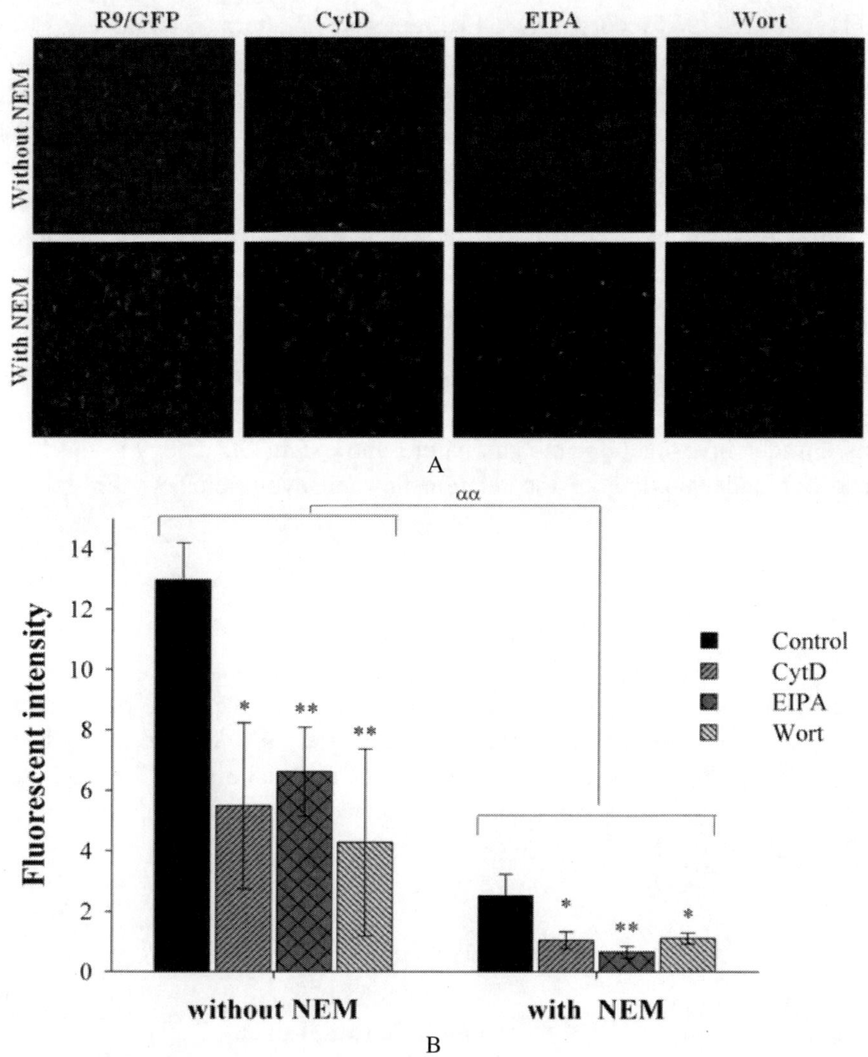

Figure 4. Effects of classical endocytic and macropinocytic inhibitors on CPP-mediated protein transduction in cyanobacteria. (A). Macropinocytic pathway. Cyanobacteria were treated with R9/GFP complexes in the absence or presence of NEM and the macropinocytic inhibitors CytD, EIPA, and Wort, as indicated. Green fluorescence was observed using a Leica TCS SL confocal microscope at a magnification of 1, 000 ×. (B). Effects of NEM and macropinocytic inhibitors on protein transduction. Results from Figure 4A were quantified using the UN-SCAN-IT software. Significant differences of $P < 0.05$ (*, α) and $P < 0.01$ (**, αα) are indicated. Data are presented as mean ± SD from five independent experiments in each treatment group.

DISCUSSION

In this chapter, we demonstrate that classical endocytosis is one of the major mechanisms for internalization of uncomplexed GFP in cyanobacteria. In contrast, both endocytosis and macropinocytosis contribute to CPP-mediated protein transduction in *Synechococcus elongatus* PCC 7942 cyanobacteria. The major mechanism for CPP-mediated protein transduction is macropinocytosis, while classical endocytosis plays a minor role. Our results are consistent with studies of bioactive macromolecular delivery in cyanobacteria [17] and eukaryotes [15, 16, 18, 19]. Accordingly, CPPs represent an alternative delivery system, other than classical endocytosis, in cyanobacteria.

Endocytosis has long been a hallmark that distinguishes eukaryotes from prokaryotes [20]. A few studies have provided evidence that prokaryotes utilize phagocytosis, pinocytosis, endocytosis, and exocytosis [3]. Unassisted macromolecule entry into prokaryotes is limited. However, we demonstrated that green (GFP) or red fluorescent proteins (RFP) noncovalently associated with CPPs can enter prokaryotes [19]. The mechanism for cellular internalization of CPP-associated protein transduction in cyanobacteria was also studied [17]. Another study with bacteria in the phylum Planctomycetes provided the evidence that GFP was internalized by endocytosis in an energy-dependent process [20]. The demonstration of endocytosis-like protein uptake in planctomycetes suggests an endomembrane system in prokaryotes [20]. Collectively, increasing evidence challenges the notion of an absence of endocytosis in prokaryotes. Further investigation into the nature and extent of endocytic machineries will contribute to our understanding of the relationship and evolution between eukaryotes and prokaryotes.

CONCLUSION

We investigated the uptake mechanisms involved in CPP-mediated protein transduction in *Synechococcus elongatus* PCC 7942 cyanobacteria. We demonstrate that classical endocytosis is one of the mechanisms for both protein uptake and CPP-mediated protein transduction in cyanobacteria. We further demonstrate that macropinocytosis plays a major and key role in CPP-mediated protein entry.

ACKNOWLEDGMENTS

We thank Dr. Hsiu-An Chu (Academia Sinica, Taipei, Taiwan) for provision of cyanobacteria and Dr. Michael B. Elowitz (California Institute of Technology, USA) for the pQE8-GFP plasmid. This work was supported by the Postdoctoral Fellowships NSC 102-2811-B-259-001 and MOST 103-2811-B-259-001 (to B.R.L.) and the Grant Number NSC 101-2320-B-259-002-MY3 (to H.-J.L.) from the Ministry of Science and Technology, Taiwan.

REFERENCES

[1] Peschek, G. A. (1984). Structure and function of respiratory membranes in cyanobacteria (blue-green algae). *Subcell. Biochem.*, 10, 85–191.

[2] Ruffing, A. M. (2011). Engineered cyanobacteria: teaching an old bug new tricks. *Bioeng. Bugs*, 2, 136–149.

[3] Herranen, M., Battchikova, N., Zhang, P., Graf, A., Sirpiö, S., Paakkarinen, V. and Aro, E. M. (2004). Towards functional proteomics of membrane protein complexes in *Synechocystis* sp. PCC 6803. *Plant Physiol.*, 134, 470–481.

[4] Jin, H., Chen, L., Wang, J. and Zhang, W. (2014). Engineering biofuel tolerance in non-native producing microorganisms. *Biotechnol. Adv.*, 32, 541–548.

[5] Chang, M., Huang, Y. W., Aronstam, R. S. and Lee, H. J. (2014). Cellular delivery of noncovalently-associated macromolecules by cell-penetrating peptides. *Curr. Pharm. Biotechnol.*, in press.

[6] Green, M. and Loewenstein, P. M. (1988). Autonomous functional domains of chemically synthesized human immunodeficiency virus tat trans-activator protein. *Cell*, 55, 1179–1188.

[7] Frankel, A. D. and Pabo, C. O. (1988). Cellular uptake of the tat protein from human immunodeficiency virus. *Cell*, 55, 1189–1193.

[8] Wadia, J. S. and Dowdy, S. F. (2002). Protein transduction technology. *Curr. Opin. Biotechnol.*, 13, 52–56.

[9] Fonseca, S. B., Pereira, M. P. and Kelley, S. O. (2009). Recent advances in the use of cell-penetrating peptides for medical and biological applications. *Adv. Drug Deliv. Rev.*, 61, 953–964.

[10] Futaki, S. (2002). Arginine-rich peptides: potential for intracellular delivery of macromolecules and the mystery of the translocation mechanisms. *Int. J. Pharm.*, 245, 1–7.

[11] Suhorustsenko, J., Oskolkov, N., Arukuusk, P., Kurrikoff, K., Eriste, E., Copolovici, D. M. and Langel, U. (2011). Cell-penetrating peptides, PepFects, show no evidence of toxicity and immunogenicity *in vitro* and *in vivo*. *Bioconjug. Chem.*, 22, 2255–2262.

[12] Liu, B. R., Huang, Y. W., Chiang, H. J. and Lee, H. J. (2013). Primary effectors in the mechanisms of transmembrane delivery of arginine-rich cell-penetrating peptides. *Adv. Stud. Biol.*, 5,11–25.

[13] Nakase, I., Takeuchi, T., Tanaka, G. and Futaki, S. (2008). Methodological and cellular aspects that govern the internalization mechanisms of arginine-rich cell-penetrating peptides. *Adv. Drug Deliv. Rev.*, 60, 598–607.

[14] Conner, S. D. and Schmid, S. L. (2003). Regulated portals of entry into the cell. *Nature*, 422, 37–44.

[15] Chang, M., Chou, J. C., Chen, C. P., Liu, B. R. and Lee, H. J. (2007). Noncovalent protein transduction in plant cells by macropinocytosis. *New Phytol.*, 174, 46–56.

[16] Liu, B. R., Huang, Y. W., Winiarz, J. G., Chiang, H. J. and Lee, H. J. (2011). Intracellular delivery of quantum dots mediated by a histidine- and arginine-rich HR9 cell-penetrating peptide through the direct membrane translocation mechanism. *Biomaterials*, 32, 3520–3537.

[17] Liu, B. R., Huang, Y. W. And Lee, H. J. (2013). Mechanistic studies of intracellular delivery of proteins by cell-penetrating peptides in cyanobacteria. *BMC Microbiol.,* 13, 57.

[18] Chang, M., Chou, J. C. and Lee, H. J. (2005). Cellular internalization of fluorescent proteins via arginine-rich intracellular delivery peptide in plant cells. *Plant Cell Physiol.*, 46, 482–488.

[19] Liu, B. R., Chou, J. C. and Lee, H. J. (2008). Cell membrane diversity in noncovalent protein transduction. *J. Membr. Biol.*, 222, 1–15.

[20] Lonhienne, T. G., Sagulenko, E., Webb, R. I., Lee, K. C., Franke, J., Devos, D. P., Nouwens, A., Carroll, B. J. and Fuerst, J. A. (2010). Endocytosis-like protein uptake in the bacterium *Gemmata obscuriglobus. Pro. Nat. Acad. Sci. USA.*, 107, 12883–12888.

INDEX

D

E

H

I

J

K

L

M

N

O

P

T

U

V

W

Y

Z